THE STORY OF SPIN

Sin-itiro Tomonaga (1906–1979)

Sin-itiro Tomonaga
THE STORY OF SPIN

Translated by Takeshi Oka

THE UNIVERSITY OF CHICAGO PRESS
CHICAGO AND LONDON

A Pacific Basin Institute Book
Major funding for the English translation of this text was provided by the Sasakawa
Peace Foundation.

The University of Chicago Press, Chicago 60637
The University of Chicago Press, Ltd., London
© 1997 by The University of Chicago
All rights reserved. Published 1997
Printed in the United States of America
19 18 17 16 15 14 13 12 11 10   5 6 7 8 9 10

ISBN-13: 978-0-226-80794-2
ISBN-10: 0-226-80794-0

Library of Congress Cataloging-in-Publication Data

Tomonaga, Shin'ichiro, 1906–79
    [Spin wa meguru. English]
    The story of spin / Sin-itiro Tomonaga ; translated by Takeshi
Oka.
        p.   cm.
    Includes bibliographical references and index.
    ISBN 0-226-80794-0
(pbk. : alk. paper)
    1. Nuclear Spin.   I. Title.
QC793.3.S6T6513   1997
539.7'25—dc21   97-12189
                                                 CIP

Spin Wa Meguru by Sin-itiro Tomonaga, copyright © 1974 by Sin-itiro Tomonaga.
English translation rights arranged with Chuokoron-sha, Inc.

⊛ The paper used in this publiction meets the minimum requirements of the
American National Standard for Information Sciences—Permanence of Pape

CONTENTS

THE STORY OF SPIN

TRANSLATOR'S PREFACE

1. Spin

This book, with the original title *Spin wa meguru*, is on the physics of spin. According to Landau and Lifshitz, "[t]his property of elementary particles is peculiar to quantum theory . . . and therefore has in principle no classical interpretation . . . In particular, it would be wholly meaningless to imagine the 'intrinsic' angular momentum of an elementary particle as being the result of its rotation 'about its own axis.' "[1] It is a mysterious beast, and yet its practical effect prevails over the whole of science. The existence of spin, and the statistics associated with it, is the most subtle and ingenious design of Nature—without it the whole universe would collapse.

Spin occupies a unique position in the teaching of physics since a wide range of physics with differing degrees of difficulty is needed for its understanding. For this reason, there has been no textbook on quantum mechanics which describes this subject in sufficient depth. Most textbooks devote at most one chapter to this subject and give only a utilitarian description. The theory of relativity, which is essential for the understanding of spin and statistics, is often forgotten except in deriving the Dirac equation. The relation between spin and statistics is apparent, yet its basis is hard to understand. Feynman wrote, "It appears to be one of the few places in physics where there is a rule which can be stated very simply, but for which no one has found a simple and easy explanation. The explanation is down deep in relativistic quantum mechanics.

1. Landau L D and Lifshitz E M 1977 *Quantum Mechanics (Non-relativistic Theory) Third Edition* p. 198 (New York: Pergamon).

This probably means that we do not have a complete understanding of the fundamental principle involved."[2]

This book is unique in that the whole of it is used for the discussion of this usually neglected subject. In his characteristically unhurried and yet rapid pace, Tomonaga cuts across the whole domain of this subject with minimal use of mathematics. This is a cozy book, and technical details are often omitted. Tomonaga's emphasis is on explaining basic ideas and their origins. This makes the book readable not only for advanced students of physics but also for general science students.

2. Contents of the Book

In lecture 1, Tomonaga starts out with a historical description of atomic spectroscopy. Three brilliant physicists, Sommerfeld, Landé, and Pauli, grope for an understanding of the multiplicity of spectra and of the Zeeman effect and find the need to introduce a new degree of freedom in addition to the orbital angular momentum of electrons. Some readers may find this lecture a little tedious since Tomonaga goes through the historical development carefully. They may skip this lecture at the first reading and return to it later.

In lecture 2, Pauli, the youngest and the most profound of the three, imaginatively assigns this new degree of freedom to the outermost "radiant" electron rather than to the inner atomic core. Two young Dutch physicists propose the audacious idea of the spinning electron. After much controversy, the idea is accepted. Even the most rigorous, Pauli, who initially rejected the idea, gives his "sanction" when Thomas explains the last persisting discrepancy of a factor of two using a relativistic transformation.

In lecture 3, Tomonaga goes right to the crux of the matter. He first discusses the introduction of the new quantum mechanics by Heisenberg and its completion by Dirac in the form of the majestic transformation theory. Pauli presents the first quantum-mechanical formulation of spin using his 2×2 matrices still based on ad hoc assumptions. Then Dirac, in a single stroke of genius (which Pauli called acrobatic), derives the relativistic equation which turns out to be the ultimate theory of electron spin. With Pauli and Dirac, the two geniuses with very different temperaments and scientific outlooks, this is one of the most exciting chapters in the book.

In lecture 4, which is allotted to the spin of the proton, the tone of the lecture changes drastically. Instead of giant physicists who moved the mainstream of physics, three molecular physicists appear—Hund, Hori, and Dennison. Their studies of molecular spectroscopy and specific heat lead to the discovery of

2. Feynman R P, Leighton R B, and Sands M 1965 *The Feynman Lectures on Physics, Vol. 3*, p. 4–3 (Reading, MA: Addison-Wesley).

proton spin. Tomonaga discusses extensively the relation between spin and statistics using the hydrogen molecule.

Lecture 5 flows back to the mainstream of physics and discusses Heisenberg's theory on the helium atom and ferromagnetism. The symmetry requirement for electron spin (Fermi statistics) results in very large, apparently magnetic, interactions. Heisenberg used this idea when he introduced isospin in nuclear physics to account for the large proton-neutron energy difference (lecture 10).

In lecture 6, Tomonaga starts to prepare for the study of relations between spin and statistics that will be discussed in lecture 8. In another acrobatic act, Dirac introduces the quantization of the field. Jordan establishes the second quantization procedure for bosons with Klein and for fermions with Wigner. It is shown that commutation and anticommutation relations are needed in order to quantize boson and fermion fields, respectively. Pauli and Weisskopf quantize the Klein-Gordon equation, together with the Maxwell equations, and show that, contrary to earlier remarks by Dirac, the equation *is* valid for bosons with spin 0, just as the Dirac equation is valid for fermions with spin 1/2. The lecture discusses the problems of the electron and the positron (Dirac equation) and Yukawa's meson (Klein-Gordon equation).

Lecture 7 is on spinor algebra, which governs the mathematical description of spin. According to Ehrenfest, "it is truly strange that absolutely no one proposed until the work of Pauli . . . and Dirac, which is twenty years after special relativity . . . this eerie report that a mysterious tribe by the name of spinor family inhabits isotropic [three-dimensional] space or the Einstein-Minkowski [four-dimensional] world." Tomonaga starts with elementary tensor algebra and proves that spinors are covariant. The algebra sets the stage for the next lecture, which is the climax of the book.

In lecture 8, Tomonaga discusses the relation between spin and statistics, the major theme of the book. As mentioned earlier, this is the most fundamental principle needed for the construction of the universe. Tomonaga first shows using the standard procedure in the theory of special relativity that for the Klein-Gordon equation the energy is single-valued and the charge is doubled-valued, while for the Dirac equation it is the opposite. He then shows, following Pauli, that these results are obtained from the general properties of tensor algebra if physical quantities are required to be covariant. Finally he shows that boson and fermion fields should be quantized through the commutation and anticommutation relations, respectively. Hence their statistics are obtained from the requirement of relativity and the covariance of physical quantities.

Lectures 9 and 10 are twin lectures in which Tomonaga discusses the establishment of nuclear physics in which the new entity *isospin* plays a major role. In lecture 9, entitled "The Year of Discovery: 1932," Tomonaga describes how the neutron was discovered by Chadwick and the positron by Anderson. These discoveries together with the timely discovery of deuterium by Urey paved the way to the application of quantum mechanics to nuclei, which once was thought to be impossible. It was realized that nuclei were composed of protons and neutrons rather than protons and electrons and that the rules of spin and statistics are applicable also to composite particles. Tomonaga shows how the properties of the neutron, i.e., its spin and magnetic moment, were determined from those of deuterium.

In lecture 10, Tomonaga first discusses the three seminal papers of Heisenberg which laid the foundation of nuclear physics. From the fact that the binding energy of a nucleus is proportional to its mass, Heisenberg concludes that the essence of the nuclear force must be an exchange force. Tomonaga explains the exchange force using the H_2^+ molecular ion. He then shows how the new concept of *isospin* was naturally introduced. He shows that the triplet splitting of the ^{14}O, ^{14}N, and ^{14}C nuclei can be explained not only qualitatively but also quantitatively using the idea of isospin analogously to the triplet state of atoms. Heisenberg's use of analogy is contrasted to Dirac's acrobatics and Pauli's frontal assault. Tomonaga then introduces Fermi's theory of β-decay and Yukawa's theory of nuclear force. He ends the lecture by noting that the problem of nuclear physics is not completely solved because of the persisting difficulty of infinity (he does not touch upon QED anywhere in the book).

Lecture 11, "The Thomas Factor Revisited," plays the role of an addendum. The late J. D. Jensen asked Tomonaga about the Thomas factor and questioned whether or not Thomas' argument merely fortuitously gave the correct result, or if it also gives the correct answer for the anomalous magnetic moments of the proton and the neutron. Tomonaga derives the magnetic moments first by the classical relativistic method of Thomas and then from the Dirac equation with Pauli's term added and shows that indeed they agree with each other. In the process he goes through the Lorentz transformation in detail.

Lecture 12, the last lecture, is composed of addenda and recollections and relates interesting anecdotes on the discovery of electron spin and Pauli's initial view of the neutrino. After talking about the Klein-Nishina formula, Tomonaga reminisces on the development of quantum mechanics and elementary particle physics in Japan and his own growth as a scientist.

3. Sin-itiro Tomonaga

When I was a high-school student in impoverished postwar Japan, the names of Hideki Yukawa and Sin-itiro Tomonaga were two lights in the darkness. They were classmates at Kyoto University, and they received the Nobel Prize for physics in 1949 and 1965, respectively, for their most fundamental works. Tomonaga was particularly influential through his textbook on quantum mechanics, the three volumes of profound books which most physics students used (the first two volumes have been translated into English and published by North Holland). His popular essays on the photon, elementary particles, symmetry, and other fundamental subjects were widely read and discussed among students. In this respect Tomonaga was like Feynman who (together with Schwinger) shared the Nobel Prize with him for the formulation of quantum electrodynamics. Tomonaga also loved the history of science and wrote the two-volume book *What Is Physics?* in which he discussed the development of classical physics from mechanics and thermodynamics to statistical physics. His description of Kepler in the book is a masterpiece. Tomonaga was not only a sharp theoretical physicist but also a very interesting human being. I recommend the recently published biography *Sin-itiro Tomonaga: Life of a Japanese Physicist* (Tokyo: MYU, 1995) for those of you who would like to know more about his person.[3]

4. Acknowledgments

This translation was initiated by Mary-Frances Jagod and Charles M. Gabrys, who were at the time graduate students in my laboratory and applied continual pressure on me to start the job and helped in recording and editing my translation. All bibliographical searching and preparation of the illustrations were done by Mary-Frances Jagod. Professor Yuzo Asano of Tsukuba University was translating the book independently and helped us greatly. The whole manuscript has been read by professors Laurie Brown, Hiroshi Ezawa, and Yochiro Nambu, whose comments greatly improved the translation. I acknowledge help from Dr. Frank Kühnemann for translating some original German text.

17 August 1996
Takeshi Oka
Enrico Fermi Institute, University of Chicago

---

3. The book may be ordered directly from MYU Publishing Company, 2-32-3 Sendagi, Bunkyo-ku, Tokyo, 113, Japan.

LECTURE ONE

Before the Dawn

Groping for the Origin of the Spectral Multiplicity

I would like to give several lectures on how spin "spins." My talks will revolve around spin, but it is also my intention, if I am successful, to talk not only about spin but also about the development of quantum mechanics so that the history of the ripening of quantum mechanics might emerge from this. Today I would like to talk about how the idea of electron spin was born and under what circumstances. The rotation of an electron about its own axis, i.e., the electron spin, was first proposed by Uhlenbeck and Goudsmit in 1925. However, there were many, many intricate developments before this idea was proposed. The story begins with the discovery of the multiplicity of spectral terms and the anomalous Zeeman effect. There was a long period of groping before the idea of electron spin was born.

As you know, Bohr published a theory for the spectrum of the hydrogen atom in 1913. According to his theory, the spectral term of the hydrogen atom can be labeled by the principal quantum number n, the subordinate quantum number k, and the magnetic quantum number m.[1] Then both n and k take integral values 1, 2, 3, . . . , and $n \geq k$. The size of the electron's orbit is determined by n, and k determines the shape; k also denotes orbital angular momentum in units of $\hbar$. (Hereafter, all angular momenta will be measured in this unit unless otherwise stated.) For a given quantum number k, the magnetic quantum number takes positive and negative integral values

$$- k \leq m \leq k, \tag{1-1}$$

---

1. Bohr used n; Sommerfeld introduced k and m.

1

and it indicates the component of the angular momentum vector **k** along the magnetic field. Inequality (1-1) limits the number of allowed m values to $2k+1$, and accordingly the number of allowed directions of the vector **k** is $2k + 1$. Probably you know that this is called *space quantization of angular momentum*. From this space quantization the Zeeman effect arises, and a single energy level will split into $2k + 1$ levels in the presence of an external magnetic field. (However, this notion will be gradually modified later.)

Now let us remember the selection rules for k and m. The rules are

$$k \quad \begin{array}{l} \nearrow \ k + 1 \\ \searrow \ k - 1, \end{array} \qquad (1\text{-}2)$$

$$m \ \begin{array}{l} \nearrow \ m + 1 \\ \rightarrow \ m \\ \searrow \ m - 1. \end{array} \qquad (1\text{-}3)$$

These mean that in a transition k changes only by ± 1 and m changes only by ± 1 or 0.

Now we can also classify spectral terms of atoms other than H by specifying n, k, and m. The reason is that, even for atoms other than H, many spectral terms correspond to an excited state of only one electron in the outermost orbit, which we shall call the *radiant electron*. (There are, of course, other states in which more than one electron are excited, but they will not be discussed here.) Now we can consider approximately that the radiant electron is moving in an electric field created by the other electrons, which we shall call the *core electrons*, and the atomic nucleus; therefore, we can tentatively regard the field to be spherically symmetric. If we can do this, then the orbit of the radiant electron can also be determined by n, k, and m, and just as in the case of hydrogen, n, k, and m are integers having the same physical meaning. One difference is that the radiant electron is buzzing around outside the core electrons, and the smallest value of n is in general greater than 1. As you know, in spectroscopy we call the term with

$$k = 1 \ \text{the} \ S \ \text{term},$$
$$k = 2 \ \text{the} \ P \ \text{term}, \qquad (1\text{-}4)$$
$$k = 3 \ \text{the} \ D \ \text{term} \ldots$$

It was soon found, however, that the terms determined by n and k are not unique, but are composed of many closely spaced levels; in other words, they have multiplet structure. For example, terms of alkali atoms, except for the S terms, are all composed of two closely spaced levels, i.e., doublets. The S term is exceptional and singlet, but as I shall explain later, even this term has potential doublet character, and if you apply a magnetic field, it splits into two levels.

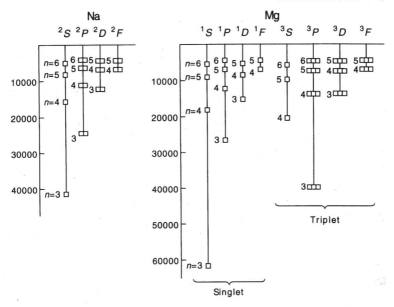

Figure 1.1 Spectral terms for alkalis (left) and alkaline earths (rights). Energy is in cm^{-1}

Therefore, we consider all the terms of alkali atoms to be doublets. In figure 1.1 I give as an example the spectral terms of Na. In this figure, the superscript 2 preceding S, P, D, and F means that they are doublet terms. We call this number the multiplicity.

I give as another example the terms of alkaline earth metals. The right-hand side of figure 1.1 gives the terms of Mg. If we look at this figure, we immediately see that the terms can be classified into two groups. In one group all the terms are singlet; in the other they are all triplet, except for the S term. In this case, the S term is singlet in both groups. It is impossible to discriminate these two from the apparent splitting, but the S term of the triplet will split into three when a magnetic field is applied. In this sense it has the potential character of a triplet, so we shall put it into the triplet group. On the contrary, splitting does not occur for the other S term even when a magnetic field is applied, and we can regard it as a bona fide singlet term. Again, the superscript on S, P, D, and F gives the multiplicity of the terms.

Thus, if there is multiplicity of the spectral terms, it is clear that the quantum numbers n, k, and m are insufficient to account for all the levels. Therefore, Sommerfeld introduced the fourth quantum number j in 1920 and used the four quantum numbers n, k, j, and m. He called this j the inner quantum number. As before, n and k determine, respectively, the size and shape of the radiant electron orbit. In addition the quantum number j was introduced to classify the variety of levels in one multiplet term, and its meaning will be considered later. Sommerfeld called it the *inner* quantum number because it classifies the levels within one multiplet term. The quantum number m now specifies the sublevels which are split for a level specified by n, j, k when the atom is in a magnetic field. This splitting is also a type of Zeeman effect, but its splitting pattern is different from that before we introduced j. Therefore this new m in the new system does not necessarily follow inequality (1-1), and both the number of sublevels and the intervals between them in the Zeeman effect are by and large different from those found previously. (I shall say later that m satisfies $-j \leq m \leq j$.)

Sommerfeld found that if j is properly chosen, the selection rule for j is

$$
j \to \begin{cases} \nearrow & j + 1 \\ & j \\ \searrow & j - 1. \end{cases}
\tag{1-5}
$$

On the other hand, for k, selection rule (1-2) holds as before. As for the quantum number m, although it differs from the previous m, experiment has shown that selection rule (1-3) still holds.

Furthermore, according to experiment, there is an additional selection rule concerning multiplicity which states that transitions occur only between terms with the same multiplicity or between terms with multiplicity differing by 2. In the case of alkali atoms, there are only doublets, and for alkaline earths there are only singlets and triplets. Therefore, this selection rule automatically holds. However, for titanium, for example, there is a quintet in addition to singlet and triplet, and vanadium has quartet and sextet in addition to doublet. In these cases transitions do not occur between singlet and quintet or between doublet and sextet.

Landé introduced as a supplementary quantum number

$$
R = \frac{\text{multiplicity}}{2}.
\tag{1-6}
$$

Therefore,

$$
\begin{aligned}
&\text{for singlets} &&R = \frac{1}{2}, \\
&\text{for doublets} &&R = 1, \\
&\text{for triplets} &&R = \frac{3}{2} \dots
\end{aligned}
\tag{1-7}
$$

Then the selection rule for the multiplicity is

$$
R \to \begin{cases} \nearrow & R + 1 \\ & R \\ \searrow & R - 1. \end{cases}
\tag{1-8}
$$

Figure 1.2 Alfred Landé (1888–1975). [Courtesy of AIP Meggars Gallery of Nobel Laureates]
Figure 1.3 Arnold J. W. Sommerfeld (1868–1951). [Courtesy of AIP Emilio Segrè Visual Archives *Physics Today* Collection]

What number should be used as the inner quantum number? We must make it such that selection rule (1-5) is followed. But j cannot be determined from (1-5) alone because even if j satisfies (1-5), you can add to it or subtract from it an arbitrary number and still it will satisfy (1-5). Between 1922 and 1925 Sommerfeld, Landé, and Pauli had been in hot competition trying to classify multiplicity and its Zeeman effect. But the three of them each used a different inner quantum number and supplementary quantum numbers.

I have already shown in figure 1.1 the spectral terms of Na and Mg. In this figure the S term looks like a singlet so I used one box ($\square$); for the P, D, and F terms, singlet is expressed by $\square$, doublet by $\square\square$, triplet by $\square\square\square$. The problem now is that for each of these boxes, we must assign a specific inner quantum number. Now Sommerfeld, Landé, and Pauli did it in three different ways. (For brevity, I shall call their conventions S*, L*, and P*, respectively.) I give in tables 1.1 and 1.2 examples of inner quantum number for each convention. In table 1.1 the numbers in the double boxes are the inner quantum numbers for the doublet alkali atoms, and in table 1.2 the numbers in single boxes and triple boxes are the inner quantum numbers for singlet and triplet alkaline earths, respectively. The values are different for each convention. I just express the inner quantum number by j in S*, J in L*, and j_p in P*. For the supplementary

Figure 1.4 Wolfgang Pauli (1900–1958). [Photograph by CERN. Courtesy of AIP Emilio Segrè Visual Archives]

Table 1.1 Inner Quantum Numbers for Alkali Doublet Terms

		^2S	^2P	^2D	^2F	
	$k =$	1	2	3	4	…
Sommerfeld	$j_a = \dfrac{\text{multiplicity} - 1}{2}$ $= \dfrac{1}{2}$	$j_a = k - 1,$				…
	$j_0 =$	0	1	2	3	
	$j =$	$\boxed{1/2}$	$1/2\ \boxed{3/2}$	$3/2\ \boxed{5/2}$	$5/2\ \boxed{7/2}$	
		$|j_a - j_0| \le j \le |j_a + j_0|,$		$-j \le m \le j$		$(1\text{-}9)_S$
Landé	$R = \dfrac{\text{multiplicity}}{2}$ $= 1$	$K = k - \dfrac{1}{2},$				…
	$K =$	$\dfrac{1}{2}$	$\dfrac{3}{2}$	$\dfrac{5}{2}$	$\dfrac{7}{2}$	
	$J =$	$\boxed{1}$	$1\ \boxed{2}$	$2\ \boxed{3}$	$3\ \boxed{4}$	
		$|K - R| + \dfrac{1}{2} \le J \le |K + R| - \dfrac{1}{2},$		$-J + \dfrac{1}{2} \le m \le J - \dfrac{1}{2}$		$(1\text{-}9)_L$
Pauli	$r = \dfrac{\text{multiplicity} + 1}{2}$ $= \dfrac{3}{2}$	$k = k,$				…
	$j_P =$	$\boxed{3/2}$	$3/2\ \boxed{5/2}$	$5/2\ \boxed{7/2}$	$7/2\ \boxed{9/2}$	
		$|k - r| + 1 \le J_P \le |k + r| - 1,$		$-j_P + 1 \le m \le j_P - 1$		$(1\text{-}9)_P$

Table 1.2(a) Inner Quantum Numbers for the Alkali Earth Singlet Term

		^1S	^1P	^1D	^1F	...
	$k =$	1	2	3	4	...
Sommerfeld $j_0 = 0$	$j_a =$	0	1	2	3	...
	$j =$	$\boxed{0}$	$\boxed{1}$	$\boxed{2}$	$\boxed{3}$	
Landé $R = \dfrac{1}{2}$	$K =$	$\dfrac{1}{2}$	$\dfrac{3}{2}$	$\dfrac{5}{2}$	$\dfrac{7}{2}$	...
	$J =$	$\boxed{1/2}$	$\boxed{3/2}$	$\boxed{5/2}$	$\boxed{7/2}$	
Pauli $r = 1$	$k =$	1	2	3	4	...
	$j_P =$	$\boxed{1}$	$\boxed{2}$	$\boxed{3}$	$\boxed{4}$	

Table 1.2(b) Inner Quantum Numbers for the Alkali Earth Triplet Term

		^3S	^3P	^3D	^3F	...
	$k =$	1	2	3	4	...
Sommerfeld $j_0 = 1$	$j_a =$	0	1	2	3	...
	$j =$	$\boxed{1}$	$\boxed{0}\,\boxed{1}\,\boxed{2}$	$\boxed{1}\,\boxed{2}\,\boxed{3}$	$\boxed{2}\,\boxed{3}\,\boxed{4}$	
Landé $R = \dfrac{3}{2}$	$K =$	$\dfrac{1}{2}$	$\dfrac{3}{2}$	$\dfrac{5}{2}$	$\dfrac{7}{2}$	...
	$J =$	$\boxed{3/2}$	$\boxed{1/2}\,\boxed{3/2}\,\boxed{5/2}$	$\boxed{3/2}\,\boxed{5/2}\,\boxed{7/2}$	$\boxed{5/2}\,\boxed{7/2}\,\boxed{9/2}$	
Pauli $r = 2$	$k =$	1	2	3	4	...
	$j_P =$	$\boxed{2}$	$\boxed{1}\,\boxed{2}\,\boxed{3}$	$\boxed{2}\,\boxed{3}\,\boxed{4}$	$\boxed{3}\,\boxed{4}\,\boxed{5}$	

quantum number concerning the multiplicity, in L* we use R as defined in (1-6), in S* $j_0 \equiv R - 1/2$, and in P* $r \equiv R + 1/2$. As for the quantum number which specifies S, P, D, F, etc., we use k as before in P*, but $j_a = k - 1$ in S*, and $K = k - 1/2$ in L*. Formulas $(1\text{-}9)_S$, $(1\text{-}9)_L$, and $(1\text{-}9)_P$ prescribe how to determine the inner quantum number when the subordinate quantum number and multiplicity are given.

$$j_a = k - 1, \quad |j_a - j_0| \le j \le |j_a + j_0|, \quad -j \le m \le j \qquad (1\text{-}9)_S$$

$$K = k - \frac{1}{2}, \quad |K - R| + \frac{1}{2} \le J \le |K + R| - \frac{1}{2}, \qquad (1\text{-}9)_L$$

$$-J + \frac{1}{2} \le m \le J - \frac{1}{2}$$

$$k = k, \quad |k - r| + 1 \le J_P \le |k + r| - 1, \quad -j_P + 1 \le m \le j_P \qquad (1\text{-}9)_P$$

As an example, let me describe S*. First we determine j_0 by using the formula $j_0 = (\text{multiplicity} - 1)/2$ on the left-hand side of (1-9)$_S$. Next we determine j_a by using $j_a = k - 1$. Then we use the inequality $|j_a - j_0| \le j \le |j_a + j_0|$ to determine the allowed values of j and finally use the last inequality $-j \le m \le j$ to determine the allowed values of m for each j. The number of m values, $2j + 1$, gives the number of split levels in the Zeeman effect. We can do the same for L* and P*, and we get the numbers in boxes in this way. The readers should try it for themselves. The relations among the quantum numbers for S*, P*, and L* are as follows

$$j = J - \frac{1}{2} = j_p - 1;$$

$$j_0 = R - \frac{1}{2} = r - 1; \tag{1-10}$$

$$j_a = K - \frac{1}{2} = k - 1.$$

All these quantum numbers satisfy selection rules, and for the magnetic quantum number you get exactly the same results regardless of the convention. The quantum numbers for singlets and triplets are given in tables 1.2(a) and 1.2(b). The prescriptions to determine the inner quantum number and magnetic quantum number are exactly the same as for doublets as given in (1-9).

Now I have given you three conventions to determine inner quantum numbers. Each has advantages and disadvantages. I shall say later that S* is followed now, but that time was one of groping, and it was very difficult to favor any one of the three. So I proceed without favoring any of them. This way, perhaps, you can better understand the tenor of that age.

I will not say any more on these conventions, but I would like to delve into the reason for the multiplicity and how people thought about this problem.

As I told you in the beginning, we assume when considering atomic spectra that the outermost electron is circling the atomic core. As long as we consider the core spherically symmetric, the energy of the electron does not depend on the tilt of the plane of the orbit unless you apply an external magnetic field. However, if the core is not spherically symmetric, e.g., if the core has angular momentum, the story is different. In this case the core has a magnetic moment associated with its angular momentum, and there will be a magnetic field which is axially symmetric about the axis of angular momentum. If this is the case, there will be an internal Zeeman effect due to this internal magnetic field. Normally, the angular momentum vector of the electron will be space-quantized with respect to this symmetry axis of the core, and for each orientation the value of the magnetic interaction between the internal magnetic field and the orbital motion will be different. This splits the term corresponding to certain n and k into multiplets. This must be the reason for the multiplets. Or so it was thought at that time.

However, the internal Zeeman effect which occurs from this differs from the Zeeman effect from an external magnetic field in one essential point. For the external Zeeman effect for the level with subordinate quantum number k, the angular momentum is quantized with $2k + 1$ values, and the term always splits into $2k + 1$ terms. However, for the internal Zeeman effect the number of split levels does not always reach $2k + 1$. This is for the following reason.

Let us take the angular momentum of the core as $\mathbf{r}$ and the orbital angular momentum of the radiant electron as $\mathbf{k}$. Then the total angular momentum $\mathbf{j} = \mathbf{r} + \mathbf{k}$, and

$$|r - k| \leq j \leq |r + k|. \tag{1-12}$$

The magnitude of the total angular momentum j can take only integral values in this range. From this the possible values of j number are

$$\begin{cases} 2r + 1 & \text{if } r \leq k \\ 2k + 1 & \text{if } r \geq k. \end{cases} \tag{1-13}$$

Therefore, the number of levels in the internal Zeeman effect, i.e., the number of levels of multiplet terms, is $2r + 1$ or $2k + 1$ depending on whether $r \leq k$ or $r \geq k$.

Next we discuss what type of Zeeman effect appears if an external magnetic field is applied to an atom that already has such multiplets. As I shall explain later, if the applied magnetic field is not that strong, then the total angular momentum j is further space-quantized along the magnetic field. In this case m, the component of j along the magnetic field, takes integral values given by

$$-j \leq m \leq j, \tag{1-14}$$

and accordingly, each level of a multiplet splits further into $2j + 1$ sublevels. We can consider this the Zeeman effect of the multiplet term.

To see whether we can explain the experimental results by this idea, let us just consider the simplest case of $r = 0$. In this case for all values of k, the first of (1-13) is valid, and the number of levels is always 1. Therefore, only a singlet term appears. It seems that we can explain the singlet term of Mg this way. Next, let us put $r = 1$. In this case also the first of (1-13) is valid for all values of k, and therefore the number of levels is always $2r + 1 = 3$. However, we cannot say that this explains the triplet terms of Mg because according to experimental results, the number of S terms in Mg is always 1, not 3. Also, for singlet terms as well as for triplet terms, the number of sublevels split by the external Zeeman effect cannot be explained by this approach.

Furthermore, if we want to explain the doublet terms of alkali atoms using (1-13), we cannot take integral values of r, so let us try $r = 1/2$. Then the first part of (1-13) is valid for any value of k, and the number of levels becomes $2r + 1 = 2$. However, the number of levels of the S term in alkali atoms is also 1, so we cannot fully explain the doublet term.

For these reasons one cannot persist in thinking that the core angular momentum is causing the multiplet terms. However, it occurred to Landé to create a substitutional model for the real atom simply to try to explain the experimental results. The inequalities in table 1.1 seem to mimic (1-12) and (1-14), which did not explain the experimental results, but somehow the entries in table 1.1 look similar, so he just made a tentative model and called it an *Ersatzmodell* ("substitutional model") and later examined how faithfully it represents the real atom.

Now we explain Landé's idea. Landé thought in his *Ersatzmodell* that the core had angular momentum **R** and its magnitude R is

$$R = \frac{\text{multiplicity}}{2}. \tag{1-15$_L$}$$

Moreover, the orbital angular momentum of the radiant electron is **K**, and its magnitude K is

$$K = k - \frac{1}{2}. \tag{1-16$_L$}$$

Needless to say, k is the subordinate quantum number. (In Sommerfeld's theory k is at the same time the orbital angular momentum, but Landé changed it slightly [to K].) If we form the total angular momentum of the system by addition of the two angular momenta, that is

$$\mathbf{J} = \mathbf{R} + \mathbf{K}, \tag{1-17$_L$}$$

and we assume for J the inequality

$$|R - K| + \frac{1}{2} \leq J \leq |R + K| - \frac{1}{2}, \tag{1-18$_L$}$$

then in this case, depending on whether $|R - K|$ is an integer or a half-integer, the allowed values of J become half-integers or integers. This prescription for angular momentum addition differs from (1-12) in that $\pm 1/2$ is on both sides of the inequality. Furthermore, Landé assumes for the component m along the magnetic field the inequality

$$- J + \frac{1}{2} \leq m \leq J - \frac{1}{2}. \qquad (1\text{-}19)_L$$

Note that again $\pm 1/2$ appears here. Instead of the unsuccessful model with k, he uses $K = k - 1/2$ [$(1\text{-}16)_L$] and also an inequality differing by $\pm 1/2$ [$(1\text{-}18)_L$] to patch up the problem. In the Landé-type prescription $(1\text{-}9)_L$ given in table 1.1 for the inner quantum number and the magnetic quantum number, he is considering this *Ersatzmodell*.

In table 1.1, in addition to Landé's prescription, I have given Sommerfeld's and Pauli's prescriptions, which we can consider as Sommerfeld's *Ersatzmodell* and Pauli's *Ersatzmodell*. For example, for Sommerfeld's *Ersatzmodell* we could consider as follows.

First we consider the core angular momentum as $\mathbf{j}_0$. Then its magnitude is

$$j_0 = \frac{\text{multiplicity} - 1}{2}, \qquad (1\text{-}15)_S$$

and if we write the orbital angular momentum as $\mathbf{j}_a$, then its magnitude is

$$j_a = k - 1. \qquad (1\text{-}16)_S$$

We can add these two angular momenta to get the total angular momentum

$$\mathbf{j} = \mathbf{j}_0 + \mathbf{j}_a. \qquad (1\text{-}17)_S$$

Then the inequality for j is

$$|j_0 - j_a| \leq j \leq |j_0 + j_a|, \qquad (1\text{-}18)_S$$

and finally, corresponding to the space-quantization of j in an external magnetic field, we have

$$- j \leq m \leq j. \qquad (1\text{-}19)_S$$

In this model Sommerfeld took for the orbital angular momentum not k but $k - 1$ in order to take care of the failure we discussed earlier. If this is done, we do not have to change the inequality $(1\text{-}12)$. Sommerfeld is the originator of space-quantization, and perhaps he did not want to change the original formula. On the other hand, Landé put $\pm 1/2$ here and there in order that the spacing rule which we shall discuss later be correctly taken care of.

You might think that Sommerfeld himself thought about this model, but that is not the case as I explain. Therefore this model is really my version of Sommerfeld's model. Actually, Sommerfeld's idea is essentially the same as the model I have just given, but he does not say that $\mathbf{j}_0$ is the core angular momentum or that $\mathbf{j}_a$ is the orbital angular momentum. If I have to tell you exactly what he says, he says as follows: "We assume the inner quantum number j to be the total angular momentum of the atom in the excited state. Then this total angular momentum is a vector sum of the unexcited atomic angular momentum $\mathbf{j}_0$ and the angular momentum of excitation $\mathbf{j}_a$ (*Impulsmoment* $\mathbf{j}_a$ *der Anregung*). As for j_a, $j_a = 0, 1, 2, \ldots$ for S term, P term, D term, etc." (What is meant by *unexcited* is not the state with minimum principal quantum number n but the state of minimum orbital angular momentum. He later replaced this word with the phrase S *state*.) He does not use $j_0 = $ (multiplicity $- 1$)$/2$, but rather he says that if he calculates multiplicity from his assumption, he finds $2 j_0 + 1$. But this is essentially the same as saying that in order to match experimental results j_0 must match this form. Therefore what he has done is exactly the same as my version of Sommerfeld's model.

You are already baptized in the new quantum mechanics. Therefore you must have immediately realized when you heard about my version of Sommerfeld's model that in the new quantum mechanics the magnitude of orbital angular momentum is not k but $\ell = k - 1$. Therefore Sommerfeld's $\mathbf{j}_a$ is nothing but the orbital angular momentum ℓ itself! Then it is quite obvious that the total angular momentum is not $\mathbf{j} = \mathbf{j}_0 + \mathbf{k}$ but $\mathbf{j} = \mathbf{j}_0 + \mathbf{j}_a$. Therefore this model is nothing but the revised model of the new quantum mechanics, which takes care of the failure I discussed earlier. However, Sommerfeld at that time did not call $\mathbf{j}_a$ orbital angular momentum but called it *angular momentum of excitation*. He may have chosen such an unnatural phrase because at that time it was generally thought that an orbit with angular momentum 0 could not be realized because the electron would hit the nucleus. Sommerfeld wrote in 1928 the wave-mechanical supplement (*Wellenmechanischer Ergänzungsband*) to his famous *Atomic Structure and Spectral Lines*, in which he derives, using the new quantum mechanics, that the size of orbital angular momentum is $\ell = k - 1$ and gives the following footnote. "In the last version of my book I introduced $j_a = k - 1$, and I discussed that in treating Landé's form (which will be discussed later) we should be using $j_a = k - 1$ instead of k. From now on, we will write this quantity ℓ rather than the inappropriate j_a."

However, as you see, it is obviously unnatural as a model to consider that the total angular momentum is obtained by vector addition of the minimum orbital angular momentum and the angular momentum of excitation. For this reason Landé rejected Sommerfeld's model.

As for Pauli's *Ersatzmodell*, it does not alter the orbital angular momentum k but adds ± 1 on the inequalities (1-12) and (1-14), trying to patch up the failure. However, as I tell you later, Pauli never really believed in models, and he is consistently trying to avoid the model-type mentality. Therefore we shall not discuss Pauli's model further.

Now what results did the *Ersatzmodell* yield? First Landé succeeded in deriving the rule for the multiplet splitting, which states that spacings between neighboring multiplet levels are in the ratio of simple integers. Now we shall plunge into this problem.

You must know that if an electron revolves in an orbit, it generates a magnetic moment. Just as we chose to measure angular momentum in units of $\hbar$, it is convenient to measure magnetic moment in units of the Bohr magneton. As you know, the Bohr magneton is

$$\text{Bohr magneton} = \frac{e\hbar}{2mc}. \tag{1-20}$$

(I remind you that e is the elementary charge and that the charge of an electron is $-e$.) If we do that, then the magnetic moment associated with the orbital angular momentum is

$$\boldsymbol{\mu}_K = -\mathbf{K}. \tag{1-21}$$

Here, of course, $\mathbf{K}$ is the orbital angular momentum in Landé's model. The core has the angular momentum $\mathbf{R}$, and that will also have a magnetic moment, which I write

$$\boldsymbol{\mu}_R = -g_0\mathbf{R}. \tag{1-22}$$

Here g_0 is undetermined, and we will simply fit its value to experimental results. In general, angular momentum and magnetic moment go together, and the ratio of the magnetic moment to the angular momentum is called the g-factor. That is,

$$g = \frac{|\text{ magnetic moment }|}{|\text{ angular momentum }|}. \tag{1-23}$$

For the orbital motion $g = 1$, and for the core $g = g_0$.

When there are two magnetic moments $\boldsymbol{\mu}_K$ and $\boldsymbol{\mu}_R$ in an atom, there should be a magnetic interaction between them, and therefore the magnetic energy is added to the usual energy of orbital motion, which is determined by n and k. This additional energy changes the atomic energy levels slightly, and we write the change W_{mag} as a constant times the inner product of $\mathbf{R}$ and $\mathbf{K}$

$$W_{\text{mag}} = \text{constant} \cdot (\mathbf{R} \cdot \mathbf{K}), \tag{1-24}$$

as if the core magnetic moment is concentrated at one point. (This constant will be discussed in the next lecture.) Now the problem is to calculate $\mathbf{R} \cdot \mathbf{K}$, i.e., to express $\mathbf{R} \cdot \mathbf{K}$ in terms of R, K, and J.

This calculation can be done right away from the definition of $\mathbf{J}$, $(1\text{-}17)_L$. We square both sides of $(1\text{-}17)_L$ and find

$$\mathbf{R} \cdot \mathbf{K} = \frac{1}{2}(J^2 - K^2 - R^2). \tag{1-25}$$

Therefore,

$$W_{mag} = \text{constant} \cdot \frac{1}{2}(J^2 - K^2 - R^2). \tag{1-26}$$

From this formula we can calculate the intervals within one multiplet. Namely, we fix K and R and take the difference between one J and its neighboring J, that is $J - 1$,

$$\Delta W_{mag} = \text{constant} \cdot \frac{1}{2}[J^2 - (J-1)^2] = \text{constant} \cdot \left(J - \frac{1}{2}\right). \tag{1-27}$$

I give in table 1.3 the ratio of the intervals obtained from this formula for the triplet terms of alkaline earths. Landé went through an enormous amount of experimental data and confirmed that this formula explains experimental results very well except for very light atoms such as He and Li.

This ends our consideration of the splitting pattern within a multiplet. But as we shall discuss in the next lecture, we can calculate not only the ratio of ΔW_{mag} between pairs of levels but the separation ΔW_{mag} itself. Actually, from this study of the energy separation between levels, Landé ended up abandoning his own idea that the reason for the multiplets is the magnetic interaction (1-24). When Landé was studying the splitting pattern, he was not really thinking of the actual magnitudes, he not being God. Rather, he was full of hope for his own model because (1-27) explains experiments so well. I shall discuss this in the next lecture.

Table 1.3 Splitting Ratios in Triplet Terms

^3S	^3P			^3D			^3F		
3/2	1/2	3/2	5/2	3/2	5/2	7/2	5/2	7/2	9/2
	| ↔ | ↔ |			| ↔ | ↔ |			| ↔ | ↔ |		
	1 : 2			2 : 3			3 : 4		

The other thing which Landé calculated from his *Ersatzmodell* is related to the Zeeman effect of the multiplet term. As I told you, the angular momenta **K** and **R** produce magnetic moments μ_K and μ_R according to (1-21) and (1-22). Thus, the atom as a whole will possess a total magnetic moment

$$\mu = -(\mathbf{K} + g_0\mathbf{R}).\tag{1-28}$$

Then the atomic energy in a magnetic field has to be the field-free energy plus

$$E_H = -\frac{e\hbar}{2mc}(\mathbf{H}\cdot\mu),\tag{1-29}$$

and therefore the atomic level changes also with magnetic field. In the limit when the external magnetic field is very much smaller or larger than the internal magnetic field, we can calculate this energy by using simple approximations.

Let us start from the weak magnetic field. We first have to realize that the total angular momentum $\mathbf{J} = \mathbf{K} + \mathbf{R}$ is a conserved quantity if there is no magnetic field. However, if the atom is in a magnetic field, the magnitude of **J** is not necessarily conserved. (In contrast, the component m along the magnetic field is conserved independently of the strength of the field.) However, if the magnetic field is sufficiently weak, J is also approximately conserved, and we shall proceed under this assumption.

First we put $g_0 = 1$. In this case

$$\mu = -(\mathbf{K} + \mathbf{R}) = -\mathbf{J}.\tag{1-28'}$$

Therefore, $\mathbf{H}\cdot\mu = -\mathbf{H}\cdot\mathbf{J} = -Hm$ if we use the magnetic quantum number. The change of energy is, from (1-29),

$$W_H = E_H = \frac{e\hbar}{2mc}Hm, \qquad -J + \frac{1}{2} \le m \le J - \frac{1}{2}.\tag{1-30}$$

(The m in $e\hbar/2mc$ is the electron mass, and you shouldn't mistake it for the magnetic quantum number m. This is confusing, but please put up with it.) In this case in the Zeeman effect, one level splits into sublevels separated by $(e\hbar/2mc)H$, and the number of sublevels is $2J$. (Note that if we use Sommerfeld's quantum number j, the number is $2j + 1$.) In general, if intervals between sublevels can be expressed in this form, we call it the normal Zeeman effect. This is the only Zeeman effect in the classical theory of Lorentz and Larmor. If we consider the case with $g_0 \ne 1$, however, a different kind of Zeeman effect will result. Such a freakish pattern is called the anomalous Zeeman effect. According to experiments, the Zeeman effects for singlet terms are all normal, and for other multiplets almost all are anomalous except for a few cases. Now we consider the case of $g_0 \neq 1$.

When $g_0 \neq 1$, (1-28') is no longer valid. In other words, $\mu \neq -J$, and therefore $H \cdot \mu$ in (1-29) is not a conserved quantity. Therefore when $g_0 \neq 1$, we cannot use (1-29) as the variation of energy W_H. When the magnetic field is small, it is known that approximately

$$W_H = -\frac{e\hbar}{2mc}(H \cdot \langle \mu \rangle). \tag{1-29'}$$

Here $\langle \mu \rangle$ is the μ averaged over time for $H = 0$. (μ, unlike J, changes with time.) If we write the component of μ along J as $\mu_\parallel$, then $\langle \mu \rangle = \langle \mu_\parallel \rangle$. If we write $\mu_\parallel$ explicitly,

$$\mu_\parallel = \frac{J \cdot \mu}{J^2} J.$$

We then write (1-28) in the form

$$\mu = -\{J + (g_0 - 1)R\}$$

and substitute this equation into the formula for $\mu_\parallel$. We immediately obtain

$$(H \cdot \langle \mu \rangle) = -\left[1 + (g_0 - 1)\frac{J \cdot R}{J^2}\right] H \cdot J.$$

Using the same argument that we used to obtain (1-25) in the discussion of multiplet splittings, we obtain $J \cdot R = (J^2 + R^2 - K^2)/2$. We have therefore

$$(H \cdot \langle \mu \rangle) = -g(H \cdot J) \tag{1-31}$$

$$g = \left[1 + (g_0 - 1)\frac{J^2 + R^2 - K^2}{2J^2}\right]. \tag{1-31'}$$

Therefore, for $g_0 \neq 1$ we can write

$$W_H = \frac{e\hbar}{2mc}Hgm, \qquad -J + \frac{1}{2} \leq m \leq J - \frac{1}{2}, \tag{1-32}$$

and the limitation on m is identical to that for (1-30). This is the formula for the anomalous Zeeman effect derived from Landé's *Ersatzmodell*. We might mention that (1-31) can be taken to mean that $|\mu|$ is proportional to $|J|$ with a ratio g. Therefore, we can consider this g as the g-factor for the entire atom.

On the other hand, Landé went through mountains of experimental data on the g-factor then available and determined the empirical formula

$$g_{exp} = \left[1 + \frac{J^2 - \frac{1}{4} + R^2 - K^2}{2\left(J^2 - \frac{1}{4}\right)} \right]. \qquad (1\text{-}31)_{exp}$$

If we compare this formula with (1-31'), it suggests that

$$g_0 = 2. \qquad (1\text{-}33)$$

However, even if we assume this value for g_0, the experimental and theoretical values do not match completely. This is because $(1\text{-}31)_{exp}$ contains the mysterious number $1/4$, which does not exist in (1-31'). Landé's *Ersatzmodell* does not give a nice rule for the g-factor as it did for the ratio of splittings within a multiplet. However, because (1-31') and $(1\text{-}31)_{exp}$ look so similar, he had great hope for his *Ersatzmodell*, so much so that Pauli warned him not to trust the model too much. I might add here that among the three models Landé's convention is the only one that gives such good agreement. In deriving (1-26) and (1-31') Landé did not use his characteristic formulas (1-6) or $(1\text{-}9)_L$ anywhere. Therefore, similar formulas can also be derived from the other conventions. For example, in Sommerfeld's convention we have instead of (1-26)

$$W_{mag} = \text{ constant } \cdot \frac{1}{2}(j^2 - j_a^2 - j_0^2), \qquad (1\text{-}26)_S$$

and we get instead of (1-31'),

$$g = 1 + (g_0 - 1)\frac{j^2 + j_0^2 - j_a^2}{2j^2}. \qquad (1\text{-}31')_S$$

The difference in the result arises from the difference in the value of the quantum numbers used here. We give in figure 1.5 the schematic pattern of the Zeeman effect for the ^2P term.

Now we go on to the case of a strong magnetic field. The experiment demonstrating the Zeeman effect under a strong magnetic field has been performed extensively by F. Paschen and E. Back, wherefore this type of Zeeman effect is often called the Paschen-Back effect. Their experiment showed that under a strong magnetic field the spacings of the split levels are all integral multiples of $(e\hbar/2mc)H$. For some multiplets this integer is 1, and the splitting resembles

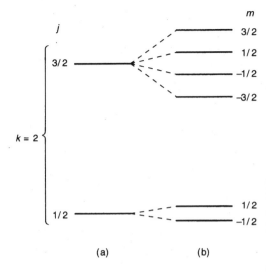

Figure 1.5 The Zeeman effect of the ^2P term. (a) Doublet levels without magnetic field. (b) Splitting of the levels due to magnetic field. The numbers beside the levels and the sublevels are the quantum numbers j and m.

the normal Zeeman effect. However, theoretically this observation is only superficial.

Modeling an atom under a strong magnetic field is much simpler than under a weak field because, if the external magnetic field is very much stronger than the internal one, then both the atomic core and the radiant electron are strongly influenced by the external magnetic field, and we can ignore the effect of the internal magnetic field. In that case, the orbital angular momentum $\mathbf{K}$ and the core angular momentum $\mathbf{R}$ will be independently space-quantized with respect to the external magnetic field $\mathbf{H}$. (We use Landé's convention here.) In other words, if we make the component of $\mathbf{K}$ along $\mathbf{H}$, m_K, and the component of $\mathbf{R}$ along $\mathbf{H}$, m_R, then m_K and m_R take integral or half-integral values satisfying

$$-K + \frac{1}{2} \leq m_K \leq K - \frac{1}{2}, \qquad -R + \frac{1}{2} \leq m_R \leq R - \frac{1}{2}. \quad (1\text{-}34)$$

It is obvious then that the magnetic quantum number m is given by

$$m = m_K + m_R. \quad (1\text{-}35)$$

(As I told you earlier, the magnitude J of $\mathbf{J}$ is not conserved if the magnetic field is very strong, but the component m along the magnetic field is conserved.) Therefore using (1-28), the component of the magnetic moment μ along the $\mathbf{H}$-axis becomes $-(m_K + g_0 m_R)$, and if we substitute this into (1-29), we immediately obtain

$$W_{\mathrm{H}} = \frac{e\hbar}{2mc} H(m_K + g_0 m_R). \qquad (1\text{-}36)$$

We can compare this formula with experiment and determine g_0. Since the interval between split levels is an integral multiple of $(e\hbar/2mc)H$, g_0 must be an integer, and from the observed splitting we see

$$g_0 = 2. \qquad (1\text{-}37)$$

In this case unlike the case of a weak magnetic field, formula (1-36) results not only from Landé's convention but also from Sommerfeld's and Pauli's conventions. (I told you earlier that for the component of angular momentum along the field, the same result is obtained regardless of the convention.) When $g_0 = 2$, agreement with experiment is perfect. So contrary to the case of a weak magnetic field where agreement is merely *suggested*, in the strong case we may consider that g_0 is definitely determined to be 2. Furthermore, (1-37) can also be confirmed from experiments under a strong magnetic field such as the Einstein–de Haas experiment. (I shall talk about this experiment later.) In figure 1.6, I show the Paschen-Back effect for a ^2P term. I show in the figure how the sublevels shift as the magnetic field is changed from the weak to the strong case. As you see in the figure, the sublevels corresponding to $j = 3/2$ and $j = 1/2$ are completely rearranged under a strong magnetic field. This clearly shows that j is not conserved under a strong magnetic field.

We have seen that many valid results came out of the *Ersatzmodell*. However, from the idea that multiplets result from the core angular momentum, many peculiar results emerge. This was pointed out by many people, including Landé. Let me explain this situation using an example. The ground state of the Na atom is ^2S as in figure 1.1, but in L* it has angular momentum $J = 1$. (See table 1.1.) On the other hand, it is experimentally known that Mg$^+$ has the same number of electrons as Na, and as a result it has spectral terms which are very similar to those of Na. (Mg$^+$ means the singly ionized, positive ion of Mg. In general, we will write the singly, doubly, triply ionized cation of atom A as A$^+$, A^{++}, A^{+++}, etc.) In fact, the ground state of Mg$^+$ is ^2S, and it should also have angular momentum $J = 1$. Now we see that if we add one electron to Mg$^+$, we get Mg atom. This means that the core of Mg atom is Mg$^+$. You might expect from this that the core of Mg atom has angular momentum 1, but as shown in table 1.2, that is not the case, and $R = 1/2$ or $3/2$. Therefore, if an electron approaches Mg$^+$ from outside and the ion becomes a core of Mg, suddenly J must change from 1 to either $1/2$ or $3/2$. Landé did not seem to worry about this situation, but Pauli deemed it extremely unnatural. For this very reason

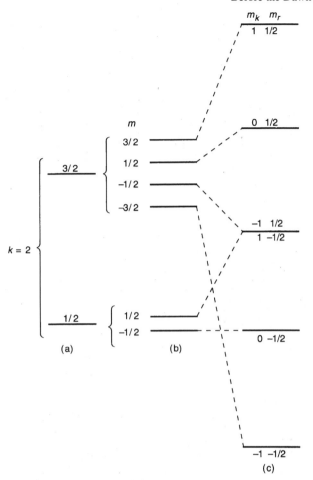

Figure 1.6 The Paschen-Back effect of the ^2P term. (a) Doublet levels without magnetic field. (b) Splitting of the levels due to magnetic field. (c) Sublevels of the Paschen-Back effect. The numbers above or below the sublevels are the quantum numbers m_k and m_r.

Pauli avoided using the *Ersatzmodell* and warned others not to place too much confidence in it.

Around that time many people noted that, if you consider the ground state of He to be composed of two electrons both in the $n = 1$ and $k = 1$ orbit, the ionization energy calculated for the He atom does not agree at all with experiment. From this discrepancy Landé hypothesized, "Out of the two electrons in He, one belongs to the core, and the other is the radiant electron, and the two electrons play entirely different roles in the atom. (E.g., the g-factor of the former is 2, and that of the latter is 1.) Therefore, both electrons cannot be in the same orbit of $n = 1$ and $k = 1$." Opposing this idea Pauli wrote in a paper submitted in October 1923 that Landé's interpretation that "the two electrons are playing entirely different roles" is the classic case of overconfidence in a model, and this precisely shows the fatal flaw of the model itself. Pauli further says that this problem of He has to be explained from an entirely different point of view. For this reason in his paper on the multiplicity, Pauli says from the beginning, "I can hardly believe the models that are currently being considered. Why don't we study the rules of multiplet terms just from experimental results?"

In fact, in the ensuing paper, Pauli consistently avoids terms like *orbital angular momentum*, *core angular momentum*, and *total angular momentum*, which imply models, and uses only *the quantum number* k, *the quantum number* r, *the quantum number* j_p, etc. Even when he had to use *j* from Sommerfeld's convention, he does not call it *total angular momentum* but *total angular momentum quantum number* and defines that as "the maximum of the magnetic quantum number *m*." (You might notice that this is in contrast to Sommerfeld, who said, "Let us assume *j* is the total angular momentum.") When the Zeeman effect is observed, sublevels appear which are symmetric with respect to the original level, and for Pauli the direct numbering of this sublevel means that *m* has the most fundamental meaning in that it directly relates to experiment. For him *j* is a more indirect thing which is defined as the maximum of *m*. This logic is the reverse of Sommerfeld's and Landé's, and it characterizes Pauli's approach that he relies just on experimental results and not on a model.

Now then, what rules did Pauli arrive at and what methods did he use? He used extremely characteristic, interesting logic, but unfortunately we have to stop here because of time. I just wish to say one thing before I end today's lecture because it is really a Pauli-like story.

It concerns whether Pauli really never thought in terms of models. That is not true. Long before Sommerfeld and Landé, Pauli also seems to have used models. In the paper submitted in October 1923, after warning of Landé's overconfidence in models, he adds another footnote. "The *Ersatzmodell* was first published by Sommerfeld and Landé in 1923, and they use it to discuss many things. However, I knew from long before [*schon seit längerer Zeit*] what would come out from the application of the *Ersatzmodell* to the behavior of atoms." Namely, Pauli already knew everything about the advantages and disadvantages of the *Ersatzmodell*. Such was Pauli's intellect. He figures out everything before other people do and studies it completely, but he never publishes it until he's absolutely convinced. He had this perfectionist tendency. Therefore, I think he really had known "from long before" the contradiction in explaining the multiplicity by the atomic core, and that convinced him that he should establish the rules only from experimental fact. In this process it seems that a new idea crystallized in his thinking around 1924, that *the origin of the multiplicity is not the core but the electron itself.* You might recall that when he warned Landé, he said, "The very fact that the two electrons in He have to play entirely different roles—one the core electron and the other the radiant electron—is the failure of the model." It seems he had already sensed that the origin of the multiplicity is in each electron itself. Pauli published this idea, and from it the idea of the spinning electron was generated by Uhlenbeck and Goudsmit in 1925. For now we finish, leaving the entrance of our hero spin to the next lecture.

Electron Spin and the Thomas Factor

The Conjecture of the Self-rotating Electron and Pauli's Sanction of It

In the previous lecture I said that the multiplicity of a spectral term is a type of internal Zeeman effect and described how the *Ersatzmodell* evolved from this idea to explain a variety of data on multiplicity. But as Pauli pointed out early on, this model in fact contains a serious problem. Furthermore, around 1924 Landé himself, who had great hope for his *Ersatzmodell*, found a fact which shook the foundation of his model. That is to say, from the hypothesis that the intervals between levels within a multiplet term are determined by a magnetic interaction of the two angular momenta (of the core and the radiant electron), Landé found only conclusions that contradicted experiments. Today I start this lecture from Landé's sad discovery.

Here we limit ourselves to the alkali atoms and similar ions. Landé pored over mountains of experimental data and found an empirical formula for the value of intervals within doublet levels. In doing this, Landé took Sommerfeld's fine structure formula, proposed in 1916, as a point of departure. Why did Landé want to use such a timeworn formula? He had a good reason, but I hesitate to present it because it takes too long. I just mention here the essence of Sommerfeld's theory.

Around 1916 Sommerfeld found that if he introduced relativity into the Bohr theory of the hydrogen atom, the relativistic variation of the mass of the electron is different for the S orbit, the P orbit, the D orbit, etc. Whereas in the nonrelativistic theory the terms belonging to the same n have the same energy, we find that once we consider relativity, the levels S, P, D, etc. will be separated. Therefore, a single term of the H atom which was thought to depend only on n becomes an assembly of several levels with fine intervals. This was Sommerfeld's idea. According to his calculation, the level splitting is

$$\Delta W_{\text{rel}} = +2 \left(\frac{e\hbar}{2mc} \right)^2 \frac{1}{a_H^3} \frac{Z^4}{n^3 k(k-1)}. \tag{2-1}$$

Since this splitting occurs in He$^+$, Li^{++}, and Be^{+++}, in addition to H, I put Z into (2-1). This equation gives the spacing between levels with subordinate quantum numbers k and $k-1$, and n of course is the principal quantum number. The Bohr radius a_H is given by

$$a_H = \frac{\hbar^2}{me^2}. \tag{2-2}$$

Formula (2-1) is called the fine structure formula, and indeed in H, He$^+$, Li^{++}, and Be^{+++} the splitting of the terms matches very well with this form.

You see from this that the theory allows for the splitting of levels with the same n and different k. This is entirely different from the splitting of levels within a doublet term of an alkali atom. (The two levels in an alkali doublet term have the same n and k, necessitating the introduction of a new quantum number j to discriminate the two.)

Landé found that in spite of this difference, the interval of two levels in an alkali doublet term can be expressed reasonably well by a formula that mimics (2-1). Namely, for a given subordinate quantum number k, the splitting of levels within a term can be reproduced very well from the formula

$$\Delta W_{\text{alkali}} = +2 \left(\frac{e\hbar}{2mc} \right)^2 \frac{1}{a_H^3} \frac{(Z-s)^4}{n^3 k(k-1)}. \tag{2-3}$$

The reason we have $Z-s$ instead of Z is that in a real alkali atom the electric field acting on the radiant electron is not eZ/r but $e(Z-s)/r$, since the nucleus is shielded by the internal electrons. From the experimental data, Landé calculated the value of s using (2-3) and found all reasonable values. For example, from the experimental values of the spacing of $n=2$, $k=2$ doublet terms, we get s values for Li, Be$^+$, B^{++}, and C^{+++} and for Na, Mg$^+$, Al^{++}, and Si^{+++} as shown in table 2.1. As you see from the table, for the Li family $s \approx 2$, and for the Na family $s \approx 9$, but this is quite understandable because for the Li family

Table 2.1

	Li	Be$^+$	B^{++}	C^{+++}		Na	Mg$^+$	Al^{++}	Si^{+++}
$\Delta W/h$	0.34	6.9	30.9	68.5	$\Delta W/h$	17.2	91.5	238	460
s	2.0	1.9	1.9	2.2	s	9.0	8.9	9.1	9.4

the number of internal electrons is 2, and for the Na family it is 10. He further examined the doublet terms of $n = 3, 4$, and 5 and found that in all cases (2-3) explains the situation fairly well. For this reason we may say that (2-3) is a good empirical formula.

Can you then derive (2-3) from the *Ersatzmodell*? In Landé's model the magnetic moment μ_R related to **R** and the magnetic moment μ_K related to **K** have a magnetic energy

$$W_{mag} = \text{constant } (\mathbf{R} \cdot \mathbf{K}),$$

and therefore the interval is given by a constant $\times (J - 1/2)$ [(1-24), (1-26), and (1-27)]. In the last lecture we did not even hint at the meaning of this constant. Now we would like to discuss it.

As I told you last time, if the core has a magnetic moment, there is an internal magnetic field with axial symmetry in the atom. Then the radiant electron in the magnetic field exhibits a type of internal Zeeman effect, which accounts for the appearance of multiplet terms. If we follow this reasoning and try to calculate the constant in the formula for W_{mag} or ΔW_{mag}, we need a complicated calculation, so it is wiser to reason as follows. Instead of the magnetic field due to the core acting on the radiant electron, we consider the magnetic field of the radiant electron acting on the core. Let us take the center of the atom as the origin and express the position of the electron by **r** and its velocity by **v**. Then according to Biot-Savart's law, a magnetic field of

$$\overset{\circ}{\mathbf{H}} = -\frac{e}{c}\frac{\mathbf{r} \times \mathbf{v}}{r^3} \tag{2-4}$$

is generated at the origin. Therefore, if we take the electronic orbital angular momentum as **K** according to Landé's convention, then $\overset{\circ}{\mathbf{H}}$ can be written as

$$\overset{\circ}{\mathbf{H}} = -\frac{e\hbar}{mc}\frac{1}{r^3}\mathbf{K}. \tag{2-4'}$$

Therefore, if the magnetic moment μ_R of the core is concentrated at the origin, then the magnetic interaction energy between $\overset{\circ}{\mathbf{H}}$ and μ_R can be written as in the normal Zeeman effect as

$$E_{mag} = -\frac{e\hbar}{2mc}(\overset{\circ}{\mathbf{H}} \cdot \mu_R). \tag{2-5}$$

[See (1-29).] We now use $\overset{\circ}{\mathbf{H}}$ in (2-4') and $\mu_R = -g_0\mathbf{R}$. [See (1-22).] We immediately obtain

$$E_{mag} = -2g_0 \left(\frac{e\hbar}{2mc}\right)^2 \frac{1}{r^3} (\mathbf{K} \cdot \mathbf{R}). \tag{2-6}$$

Here $1/r^3$ varies with time, and therefore we cannot immediately take this as W_{mag} of (1-24). If we take the time average of $1/r^3$ as $\langle 1/r^3 \rangle$, however, then W_{mag} can be approximated as

$$W_{mag} = -2g_0 \left(\frac{e\hbar}{2mc}\right)^2 \left\langle \frac{1}{r^3} \right\rangle (\mathbf{K} \cdot \mathbf{R}). \tag{2-6'}$$

and therefore we obtain

$$\Delta W_{mag} = -2g_0 \left(\frac{e\hbar}{2mc}\right)^2 \left\langle \frac{1}{r^3} \right\rangle \left(J - \frac{1}{2}\right). \tag{2-6''}$$

So we now know the meaning of the constant, but in order to calculate the interval using (2-6"), we need to calculate $\langle 1/r^3 \rangle$. This average can be calculated exactly when an electron moves in a Coulomb field. If we set the potential equal to

$$V(r) = -\frac{Ze}{r}, \tag{2-7}$$

then according to Bohr's theory,

$$\left\langle \frac{a_H^3}{r^3} \right\rangle = \frac{Z^3}{n^3 k^3}. \tag{2-8}$$

Here, n is the principal quantum number, k is the subordinate quantum number (previously the orbital angular momentum), and a_H is the Bohr radius (2-2). If we are to use Landé's *Ersatzmodell* instead of Bohr's theory, then since the orbital angular momentum is $\mathbf{K}$, it is more consistent to use instead of (2-8)

$$\left\langle \frac{a_H^3}{r^3} \right\rangle = \frac{Z^3}{n^3 K^3}. \tag{2-8'}$$

Now we limit ourselves to the case of an alkali doublet. In this case we can replace $J - 1/2$ in (2-6") with K. Therefore, we obtain

$$\Delta W_{mag} = -2g_0 \left(\frac{e\hbar}{2mc}\right)^2 \frac{1}{a_H^3} \frac{Z^3}{n^3 K^2}. \tag{2-6$_L$}$$

Furthermore, in the calculation of the g-factor which I discussed in the previous lecture, we had to replace J^2 in (1-31') with $J^2 - 1/4$ [(1-31)$_{exp}$]. Therefore it might be better to use $K^2 - 1/4$ instead of K^2. Then $K^2 - 1/4 = (K + 1/2)(K - 1/2)$, and since Landé's K is related to k by $K = k - 1/2$, we should use $K^2 - 1/4 = k(k - 1)$. Doing this, we finally get the formula

$$\Delta W'_{mag} = -2g_0 \left(\frac{e\hbar}{2mc}\right)^2 \frac{1}{a_H^3} \frac{Z^3}{n^3 k(k-1)}. \qquad (2\text{-}6')_L$$

Now this formula is derived on the assumption that the electron is moving in a Coulomb field. Therefore, in order to generalize this formula to all atoms, we should correct for the screening of the core by changing Z to $Z - s$. We thus obtain the conclusion

$$\Delta W'_{mag} = -2g_0 \left(\frac{e\hbar}{2mc}\right)^2 \frac{1}{a_H^3} \frac{(Z-s)^3}{n^3 k(k-1)}. \qquad (2\text{-}6'')_L$$

Thus we find from Landé's idea that the interval of the levels is given by (2-6")$_L$. Comparing this result with the empirical formula (2-3), we see that they resemble one another, but there are still major differences:

  (i) $(Z - s)^4$ in (2-3) is $(Z - s)^3$ in (2-6")$_L$;
  (ii) the $+$ in (2-3) is a $-$ in (2-6")$_L$; and
  (iii) the factor of 2 in (2-3) is $2g_0$, i.e., 4, in (2-6")$_L$.

Of these three differences, the first is fatal, and if you determine s from experimental results using (2-6"), you get entirely different values from those listed in table 2.1. Thus Landé was forced to admit the failure of his idea that multiplet terms are caused by magnetic interaction. I might add that Landé seems to have missed discrepancies (ii) and (iii). He always focuses on the "interval" without worrying much about the sign. Also, he dropped $g_0 = 2$ by mistake in the middle of the calculation.

Actually, around 1922, before Landé calculated (2-6")$_L$ using the *Ersatzmodell*, Heisenberg had calculated the interval of Li doublet levels in his own way. His model resembles Landé's model, and he reported that his calculated results agreed well with experimental values. A man by the name of D. Roschdestwenski had also done a similar calculation, but the reason both he and Heisenberg found a value that agrees with experiment is that the value of Z was so small.

After these endeavors Landé wrote his last paper on this subject in April 1924 and concluded with the following sentence: "An explanation of the X-ray- and optical doublet–level intervals using magnetic forces is impossible once and for all, in spite of the success of Roschdestwenski and Heisenberg on Li." He sounds so nonchalant. After struggling so much, Landé may have reached nirvana and had no other recourse than just to accept fact as fact.

In the meantime, what was Pauli up to? After writing the paper submitted in October 1923 in which he warned Landé as I told you earlier, he did not write any papers for one year. Probably during that time he was inching toward the right solution to these very important problems related to the core, and only in December 1924 did he finally finish another paper. In this paper he unveiled his criticism of the prevailing assumption that the atomic core is in the K shell, and he took the plunge, insisting for the first time that the multiplet is due not to an interaction between the core and the radiant electron but to a characteristic of the *electron itself.*

While the *Ersatzmodell* was still popular around 1923, many people believed that the core is in the K shell. When I discussed the core last time, I included in the core all the electrons other than the outermost radiant electron. However, people thought that of these electrons, only those in the K shell have angular momentum $\mathbf{R}$ and $g_0 = 2$. All the other electrons in the other closed shells, like the L shell, the M shell, etc., were not thought to contribute to the angular momentum or magnetic moment. Pauli began by criticizing this idea.

Pauli argued that if indeed only the K shell electrons form the core, then those electrons will move quickly for large Z values and thus are expected to be greatly affected by relativity, resulting in large relativistic corrections to the ratio |magnetic moment|/|angular momentum| . He quoted Sommerfeld's paper, calculated this correction, and found that for atoms with large Z the value of g_0 should show an experimentally observable deviation from 2. Nevertheless, in reality $g_0 = 2$ regardless of the Z value. Therefore, the idea that the core is in the K shell is suspect. This is Pauli's criticism. Pauli further insisted that it was unnatural to assume that only the K shell had an angular momentum and a magnetic moment and that no other closed shells did. He insisted rather that the K shell, like any other closed shell, could have neither angular momentum nor a magnetic moment.

In addition to these reasons, Pauli also counted as detrimental Landé's unnatural description of the ground state of He (I have already mentioned that Pauli warned Landé) and the failure to explain the level intervals within a multiplet as a magnetic interaction between the core and the radiant electron (I talked about this earlier), and he concluded as follows:[1]

> The closed electron configurations shall not contribute to the magnetic moment and the angular momentum of the atom. Especially in the case of the alkalis, the value of the angular momentum of the atom and its change in energy in an external magnetic field are considered to be mainly due to the radiant electron, which is also thought to be the origin of the magneto-mechanic anomaly. From this point of view the doublet structure of the

1. Tomonaga made a loose German-to-Japanese translation of this passage. We have substituted for it a direct German-to-English translation.

alkali spectra, as well as the breakdown of the Larmor theorem, is caused by a strange two-valuedness of the quantum-theoretical properties of the radiant electron which cannot be described classically.

Here, Pauli for the first time declared that of the four quantum numbers n, k, j, and m, j and therefore m do not arise from an interaction between the radiant electron and the core, but all belong to the electron itself. However, even here he tried to avoid the image of a model behind these quantum numbers and simply said "classically indescribable two-valuedness."

To justify this assertion, Pauli tried to use it to interpret a variety of things. One of them is that the "peculiarity" plaguing the idea of the core is completely eliminated using his idea. As I have already said many times, if you adopt the core concept, some property of the core must suddenly change when an electron is added to the core. This absurdity does not appear in Pauli's new way of thinking. Furthermore, in his new idea all the electrons have "classically indescribable two-valuedness," and therefore there is no difference between the core electrons and the other electrons.

Let me explain the "peculiarity," sticking as much as possible to Pauli's words. Let us consider an alkali atom. If we observe the Zeeman effect of this atom in a weak magnetic field and count the number of sublevels, one level of the doublet term splits into $2(k-1)$ sublevels, and the other splits into $2k$ sublevels except for the S term. Therefore, if we add the two levels, the total number of sublevels will be $2(2k-1)$. Since the S term is singlet, it constitutes an exception, but since it splits into two levels in the Zeeman effect, the number $2(2k-1)$ is also valid here. For this reason, for a given n and k of an alkali atom, the number of states belonging to n, k is $N_2(k)$,

$$N_2(k) = 2(2k-1) \qquad\qquad (2\text{-}9)_2$$

(the subscript 2 means doublet). We can further derive this number for the Paschen-Back effect in a strong magnetic field or for an intermediate magnetic field. For the special case of $k = 1$, the S term,

$$N_2(1) = 2. \qquad\qquad (2\text{-}9')_2$$

Next let us consider alkaline earth atoms. In this case, there arise singlet and triplet terms, and from observations of the Zeeman and Paschen-Back effects, we obtain

$$N_1(k) = 1(2k-1), \qquad\qquad (2\text{-}10)_1$$
$$N_3(k) = 3(2k-1). \qquad\qquad (2\text{-}10)_3$$

The subscripts 1 and 3 on N denote singlet and triplet terms, respectively. Therefore, if we do not discriminate between singlet and triplet and give only n and k, the number of states is

$$N_1(k) + N_3(k) = 4(2k - 1). \qquad (2\text{-}10)_{1+3}$$

The result and conclusions presented so far were obtained simply by sorting out experimental facts, and we have not used any model. Now let us consider what will emerge if we interpret these conclusions using the *Ersatzmodell*.

As I told you last time, the explanation of the Paschen-Back effect using the *Ersatzmodell* was quite straightforward. In this case, because of the strong magnetic field, the magnetic component of the core angular momentum **r** and the orbital angular momentum **k** are individually space-quantized. As a result, in alkali doublets, the magnetic component m_r of **r** takes two values, and the magnetic component m_k of **k** takes $2k - 1$ values (since we are now using Pauli's convention, $-r + 1 \leq m_r \leq r - 1$ and $-k + 1 \leq m_k \leq k - 1$). Therefore for a given k, the number of states is $2(2k - 1)$. From this consideration we might say that the first factor 2 of $(2\text{-}9)_2$ is the number of states of the core and the second factor is the number of states of the radiant electron.

The same thing can be said about the alkaline earths. Namely, the first factors 1 of $(2\text{-}10)_1$ and 3 of $(2\text{-}10)_3$ represent the number of core states for the singlet term and the triplet term, respectively. And for both multiplet terms the number of states of the radiant electron is the same as in the alkali atom, $2k - 1$.

Now the "peculiarity" appears. For example, let us consider Mg^+. As I said already many times, Mg^+ must have a term similar to that found in Na, and therefore its ground state is a doublet according to $(2\text{-}9')_2$. Therefore, if a radiant electron with quantum number k is added to the ion to form Mg atom, since the radiant electron always has $2k - 1$ states, the total number of states has to be $2(2k - 1)$. However, this is not the case, and the actual number of states of Mg is $1(2k - 1)$ or $3(2k - 1)$. For this reason, the number of states, 2, of Mg^+ has to change abruptly from 2 to 1 or 3 at the moment when the electron joins it. This is indeed "peculiar." There is no way to explain this with classical mechanics and the correspondence principle. Bohr therefore said that this occurs through "nonmechanical constraint" (*nichtmechanischer Zwang*).[2]

Pauli insists here that if people adopt his new idea that the electron itself has two-valuedness, then there is no need for such an unnecessary complication as nonmechanical constraint. Specifically, he argues that this "peculiar" situation results from the idea that the factors 2 and $2k - 1$ come from the number of states of the core and the number of states of the radiant electron, respectively.

---

2. Various historians of physics have translated *Zwang* as "coercion," "constraint," and "force." We have adopted "constraint," which is used in the English translation of this paper in *Niels Bohr Collected Works*.

But if you consider $2(2k - 1)$ as the number of states of one electron, then this awkwardness does not occur. We simply say that when an electron attaches to Mg^+, the number of states of Mg^+ is 2 and the number of states of the attached electron is $2(2k - 1)$, and therefore the number of states for a given k is

$$2 \cdot 2(2k - 1). \tag{2-10}$$

Thus we obtain (2-10) $_{1+3}$ without introducing any peculiar idea. Why then must the number of states of an electron be $2(2k - 1)$? Pauli calls the first factor 2 a "classically indescribable two-valuedness" and does not want to go further into the problem.

According to Pauli, this new concept is of course still incomplete, and you cannot explain why the factor 4 in (2-10) $_{1+3}$ becomes 1 and 3—in other words, why there are two sorts of term series, singlet and triplet, how the difference of singlet and triplet arises, and why the level interval of singlet and triplet results. However, if we adopt this idea, we can explain the remarkable fact, well known experimentally, that each orbit in an atom has a certain capacity, and the observation that an orbit is closed when its capacity is filled can be explained by a very clear rule: Once we specify the values of the quantum numbers n, k, j, and m, then not more than one electron with those quantum numbers can exist within an atom. This rule can be interpreted as saying that once an electron enters into a state specified by n, k, j, and m, that electron prevents the other electrons from entering the same state, and in that sense it has come to be called *Pauli's exclusion principle*. From this rule, Pauli showed that it is possible to derive the result that the K shell closes with 2 electrons, the L shell closes with 8 electrons, and the M shell closes with 18 electrons, giving strong support to his new theory. The aforementioned rule is possible only when the n, k, j, and m are related to one electron.

Around that time it was often observed that some terms which were expected to be present were missing. For example, look at figure 1.1; the singlet terms of Mg start from $n = 3$, but the triplet terms start from $n = 4$, with $n = 3$ missing. Pauli showed this phenomenon can also be explained by the exclusion principle. Furthermore, if you consider larger atoms or atoms in which more than one electron is excited, then many levels should be missing because of the exclusion principle. In order to judge the validity of Pauli's idea, we should consider all cases and confirm that there is no discrepancy. Therefore, Pauli traveled to Landé's lab and tried to find among the many data Landé had accumulated any case that broke this rule. In Landé's laboratory he met a young man who wanted to propose that the electron is self-rotating as an extension of Pauli's idea.

That young person was R. de L. Kronig, who conceived of the self-rotating electron half a year earlier than Uhlenbeck and Goudsmit. Kronig was barely twenty years old at that time, but he was interested in multiplicity and the Zeeman effect and came from the United States to Landé's lab in January 1925. Landé showed him a letter from Pauli in which "the classically indescribable two-valuedness" was mentioned. After reading this, Kronig immediately thought of a self-rotating electron, i.e., an electron rotating about its own axis with an angular momentum of self-rotation of $1/2$ and a g-factor of $g_0 = 2$. From this assumption he could explain the appearance of the alkali doublets, the Zeeman effect, and the Paschen-Back effect as well as he could from the *Ersatzmodell*. Furthermore, he anticipated that the interaction between the magnetic moment of the self-rotation and the orbital motion could be derived through relativity, and using that, he was able to calculate the interval within doublet terms.

How do we use relativity here? Let me explain. The electrons in an atom are in an electric field and are therefore influenced by that electric field, but the self-rotating magnetic moment of the electron will not be directly affected by

Figure 2.1 From left: Yoshio Nishina (1890–1951), David M. Dennison (1900–1976), Werner Kuhn (1899–1963), Ralph de Laer Kronig (1904–1995), Bidu Bhusan (B.B.) Ray in Copenhagen, 1925.

the electric field. However, if an electron is moving, then a magnetic field which did not exist in the laboratory frame appears in the rest frame of the electron through a Lorentz transformation. According to Einstein, this field is

$$\overset{\circ}{\mathbf{H}} = \frac{1}{c} \frac{\mathbf{E} \times \mathbf{v}}{\sqrt{1 - v^2/c^2}}. \tag{2-11}$$

In this event if we write the self-rotating magnetic moment as $\boldsymbol{\mu}_e$, then its interaction with $\overset{\circ}{\mathbf{H}}$ is

$$E_{\text{rel}} = -\frac{e\hbar}{2mc}(\overset{\circ}{\mathbf{H}} \cdot \boldsymbol{\mu}_e) \tag{2-12}$$

$$= g_0 \frac{e\hbar}{2mc}(\overset{\circ}{\mathbf{H}} \cdot \mathbf{s}).$$

[Remember (2-5) of the *Ersatzmodell*.] This is the interaction energy between the magnetic moment due to self-rotation and the orbital motion. However, since (2-12) is the energy of the electron in the rest system, the energy in the laboratory system has to be obtained by a Lorentz transformation of (2-12). For this purpose, we simply multiply by $1/\sqrt{1 - v^2/c^2}$.

However, if the electric field $\mathbf{E}$ is a Coulomb field and if $v^2/c^2 \ll 1$, then we can do this calculation by a more elementary method. This is because we can obtain the magnetic field $\overset{\circ}{\mathbf{H}}$ for the electron rest system by using Biot-Savart's law. Observers sitting on top of the electron and moving with it will think the electron is at rest and instead the nucleus is revolving around the electron. They will say the nuclear velocity is $-\mathbf{v}$, the radius is $-\mathbf{r}$, and its charge is $+Ze$. Therefore, according to Biot-Savart's law, the magnetic field at the position of the electron is given by

$$\overset{\circ}{\mathbf{H}} = +\frac{Ze}{c} \frac{\mathbf{r} \times \mathbf{v}}{r^3}. \tag{2-11'}$$

[Let me remind you that the magnetic field at the core was given by (2-4) in the core model. There we had a rotating electron with charge $-e$, but here we have a nucleus with charge $+Ze$.] Now if we substitute into (2-11) the Coulomb field

$$\mathbf{E} = \frac{Ze}{r^3} \mathbf{r}$$

and approximate $\sqrt{1 - v^2/c^2} \approx 1$, then we obtain $\overset{\circ}{\mathbf{H}}$, which was given in (2-11'). If we compare (2-11') thus obtained with (2-4), which we used for the core model, we have Ze instead of e and $+$ instead of $-$. For this reason if we calculate using Kronig's idea of a self-rotating electron, then the level interval is

Z times the value given by $(2\text{-}6')_L$, and the $-$ should be changed to $+$. Namely, we have

$$\Delta W_{\text{rel}} = +2g_0 \left(\frac{e\hbar}{2mc}\right)^2 \frac{1}{a_H^3} \frac{Z^4}{n^3 k(k-1)}. \tag{2-13}$$

If the electric field is not a Coulomb field, we can correct for screening and obtain

$$\Delta W_{\text{rel}} = +2g_0 \left(\frac{e\hbar}{2mc}\right)^2 \frac{1}{a_H^3} \frac{(Z-s)^4}{n^3 k(k-1)}. \tag{2-13'}$$

If we compare this formula with Landé's empirical formula (2-3), they agree much better than in the case of the core model. We summarized the differences between experiment and the core model as points (i), (ii), and (iii), but the problems of (i) and (ii) are already resolved. However, the third discrepancy remains. As I told you before, the factor 4 results because $g_0 = 2$ was used, but we cannot put $g_0 = 1$ because neither the anomalous Zeeman effect nor the Paschen-Back effect would result. For this reason, Kronig's anticipation that the correct level interval will come out from the relativistic energy (2-12) was not realized. However, have we not done quite well to have eliminated discrepancies (i) and (ii)?

Kronig knew long before carrying out the detailed calculation that if the magnetic field in (2-11) is used, the level interval is proportional to Z^4 and not Z^3, and therefore the fine structure of H, He$^+$, Li^{++}, Be^{+++} is better explained by the self-rotation of the electron than by Sommerfeld's model. In fact, only if the factor on the right-hand side of (2-13) is 2 does Kronig's formula have precisely the same form as Sommerfeld's fine structure formula (2-1). Therefore, the fine structure can be explained by his new interpretation. Nevertheless, this prospect was not realized because of the factor 4 in (2-13).

Kronig talked to Landé about his result. Landé said that since Pauli is coming here soon, why not ask his opinion? Pauli and Kronig met, and when Pauli heard Kronig's idea, he showed no interest in it and greatly disappointed Kronig by his coolness. Kronig then went to Copenhagen and aired his idea there, but still he did not gain much sympathy. Also, he himself was not confident because of the difference from experiment by a factor of two in the level intervals, and furthermore, the idea of a self-rotating electron presents a variety of difficulties when examined within the framework of classical electron theory. (For example, if the size of the electron is e^2/mc^2 as H. A. Lorentz has considered, then so fast a rotation is needed to have a self-rotating angular momentum of $1/2$ that the electron's surface reaches a speed ten times higher than that of light.) For all of these reasons, Kronig decided not to publish his idea.

Meanwhile in the fall of 1925, Uhlenbeck and Goudsmit published precisely the same idea in *Naturwissenschaften*. Actually, after submitting their paper, they asked the opinion of Lorentz, who said it was almost impossible in classical electron theory. So an episode followed in which they tried to withdraw their paper in a hurry, but they were too late. For better or for worse, their paper was published in the journal.

When they published this paper, they had not calculated the level intervals of doublet states. However, during the mere half year after Kronig had aired his idea in Copenhagen, the atmosphere there had drastically changed. Bohr began to show much interest in the idea of the daring duo, Einstein advised them to use (2-11), which Kronig already used to calculate the level interval, and Bohr even appended a short recommendation to their paper. However, Pauli said in his Nobel lecture that he was still definitely opposed to Uhlenbeck and Goudsmit's idea. Now as for Kronig, he sent a letter to *Nature*, where their second paper was published, criticizing their paper and enumerating many difficulties of a self-rotating electron. At the end of the paper, he wrote, "The new hypothesis,

Figure 2.2 Oscar Klein (1894–1977), George E. Uhlenbeck (1900–1988), and Samuel A. Goudsmit (1902–1978), 1926. [Photograph by H. Knauss. Courtesy of AIP Emilio Segrè Visual Archives]

therefore, appears rather to effect the removal of the family ghost from the basement to the sub-basement instead of expelling it definitely from the house." Hard-nosed, is he not?

Uhlenbeck and Goudsmit, however, were well aware of a variety of problems with the self-rotating electron. (It was for exactly this reason that they wanted to withdraw their paper from *Naturwissenschaften*.) In the second paper they pointed out that when they calculated the level interval for the doublet term, there was a factor of 2 discrepancy with experiment.

While the proposition of the self-rotating electron was thus in turmoil, the famous work of L. H. Thomas appeared. In this work he showed that the discrepancy between the theory and experiment of the level interval of the doublet state is caused by an incorrect definition of the electron rest system. When this was clarified, even Pauli decided not to categorically oppose the idea of the self-rotating electron, and in spite of its remaining problem, he approved this idea. (As I told you earlier, Pauli's perfectionism was phenomenal. He not only wanted to be perfect himself, but was well known for applying this standard to other people, sometimes arguing harshly against other people's work. For this reason when people asked his opinion of their work and he gave his assent, they called it Pauli's *sanction*.)

Thomas' calculation is awfully complicated, really too much to handle here. In essence it says that "the coordinate system in which the electron is at rest and the nucleus is rotating" should not be treated so simply. If the electron is moving with constant velocity, then such a coordinate system can easily be obtained from the laboratory coordinate system by a Lorentz transformation. However, if the electron has acceleration, there is a complication. We can indeed consider the coordinate system in which the electron is motionless. Specifically, this is a coordinate system which has the electron at the origin and is moving together with the electron. However, we cannot uniquely determine the coordinate system just by declaring that its origin is moving with the electron. How should we orient the x, y, and z axes? In addition to the condition that the origin is moving with the electron, we must add the condition that the x, y, and z axes are always moving in translation, i.e., not rotating. This "parallel motion" is obvious when the electron is maintaining a constant velocity, but when the electron has acceleration, it is not that obvious. This is the point that Thomas realized.

Therefore Thomas first discussed what is meant by the parallel translation of the coordinate axes. He concluded that the parallel translation of the axes means that the axes at each instant are parallel to the axes at an infinitesimally small time before that instant. The coordinate system whose origin is moving with the electron, in this sense translating parallel to itself, may be called the proper coordinate system of the electron; Thomas derived that this coordinate system, seen from the laboratory system, is not translating parallel to itself but is accompanied by rotation if the electron has acceleration. After a laborious calculation he found that the axes of the proper coordinate system are rotating with respect to the laboratory-fixed system with an angular velocity $\boldsymbol{\Omega}$.

$$\boldsymbol{\Omega} = \frac{1}{2c^2} \mathbf{a} \times \mathbf{v}. \tag{2-14}$$

Here $\mathbf{a}$ is the acceleration of the electron. (We have used the approximation $1 - v^2/c^2 \approx 1$.) Thomas pointed out that in the calculations by Kronig and by Uhlenbeck and Goudsmit, this rotation of the proper coordinate system is not taken into account, and therefore their answer is correct if and only if the acceleration of the electron is 0. I am going to talk about Thomas' theory in detail later (lecture 11), but let me outline it here.

First of all, in the rest system of the electron we have a magnetic field $\overset{\circ}{\mathbf{H}}$ given by (2-11). In this respect Kronig and Uhlenbeck and Goudsmit were correct (and so of course was Einstein). Therefore, the energy of the self-rotating magnetic moment μ_e of the electron in this magnetic field is given by (2-12), and a sort of internal Zeeman effect occurs. Now Thomas' theory is purely classical and, like the idea of J. Larmor, explains the Zeeman effect based on the precession of μ_e. Let us first consider the precession of μ_e in the magnetic field $\overset{\circ}{\mathbf{H}}$ using the classical theory of tops. The angular velocity of precession is then

$$\boldsymbol{\Omega}_{\overset{\circ}{\mathbf{H}}} = g_0 \frac{e}{2mc} \overset{\circ}{\mathbf{H}}. \tag{2-15}$$

Here I have to digress a little bit because you may worry about the relation between this formula and (2-12). First, the fundamental frequency $\nu_{\overset{\circ}{\mathbf{H}}}$ of the precession is given by the time average of (2-15) divided by 2π, namely

$$\nu_{\overset{\circ}{\mathbf{H}}} = g_0 \frac{e}{4\pi mc} \langle \overset{\circ}{\mathbf{H}} \rangle.$$

Next, using the correspondence principle, Bohr's relation $h\nu = |W_1 - W_2|$, means that

$$h\nu_{\overset{\circ}{\mathbf{H}}} = g_0 \frac{e\hbar}{2mc} \langle \overset{\circ}{\mathbf{H}} \rangle \tag{2-15'}$$

should agree with the level interval of the doublet term calculated from (2-12). Now in (2-12), $\mu_e = -g_0 s$, and the component of s along the magnetic field is either $+1/2$ or $-1/2$. Therefore, the level interval obtained from (2-12) is

$$\Delta W_{\text{rel}} = g_0 \frac{e\hbar}{2mc} \langle \overset{\circ}{\mathbf{H}} \rangle, \tag{2-15''}$$

which indeed agrees with that given in (2-15').

Now you must have a clear picture. Speaking in terms of classical theory, what Kronig and Uhlenbeck and Goudsmit had done was to think that the angular velocity of (2-15) is itself the angular velocity of the precession in the laboratory system. However, Thomas argued as follows. This angular velocity of precession is indeed the angular velocity as seen from the proper coordinate system of the electron, but if you view the motion from the laboratory system, the proper coordinate system is itself rotating with an angular velocity given by (2-14). Therefore if we express the precession of μ_e seen from the laboratory coordinate system as Ω_{lab}, then

$$\Omega_{lab} = \Omega_{\overset{\circ}{H}} + \Omega \tag{2-16}$$

$$= g_0 \frac{e}{2mc} \overset{\circ}{H} + \frac{1}{2c^2} a \times v.$$

Now the acceleration of the electron a is related to the electric field by

$$a = -\frac{e}{m} E. \tag{2-17}$$

Therefore, we substitute this into the second term on the right-hand side of (2-16) and, using (2-11), obtain

$$\Omega_{lab} = (g_0 - 1) \frac{e}{2mc} \overset{\circ}{H}. \tag{2-18}$$

[We put $1 - v^2/c^2 \approx 1$ in (2-11).] If we substitute $g_0 = 2$, then we obtain one-half of the result of (2-15), which we derived without considering Ω. Therefore, again using the correspondence principle, we obtain, instead of the level-spacing of (2-15") which was directly obtained from (2-12),

$$\Delta W_{rel} = (g_0 - 1) \frac{e\hbar}{2mc} \langle \overset{\circ}{H} \rangle. \tag{2-19}$$

This means that as seen from the laboratory system, the interaction energy between μ_e and $\overset{\circ}{H}$ is given not by (2-12) but by

$$E_{rel} = (g_0 - 1) \frac{e\hbar}{2mc} (\overset{\circ}{H} \cdot s), \tag{2-19'}$$

and if E is a Coulomb field, then instead of (2-13) we derive

$$\Delta W_{rel} = +2(g_0 - 1) \left(\frac{e\hbar}{2mc} \right)^2 \frac{1}{a_H^3} \frac{Z^4}{n^3 k(k-1)}, \tag{2-20}$$

and for a screened nucleus, instead of (2-13') we put

$$\Delta W_{\text{rel}} = +2(g_0 - 1) \left(\frac{e\hbar}{2mc} \right)^2 \frac{1}{a_{\text{H}}^3} \frac{(Z - s)^4}{n^3 k(k - 1)}, \qquad (2\text{-}20')$$

If we substitute $g_0 = 2$, we indeed get a level spacing which is half that obtained from (2-13) or (2-13'). Therefore, out of the three problems (i), (ii), and (iii), the only remaining discrepancy, (iii), has also been resolved.

Since we obtain a result multiplying (2-12) by $1/2$ if we use $g_0 = 2$ in (2-19'), we call this factor of $1/2$ the *Thomas factor*. However, we might note that the factor $1/2$ results only when $g_0 = 2$. If $g_0 \neq 2$, we cannot say that (2-19') is half of (2-12). Therefore, I do not like to say, "The Thomas factor is $1/2$," because then we may be led to the misunderstanding that we simply have to take half of the answer obtained without Thomas' insight in order to incorporate his result.

Finally, I wish to add the following. So far we have considered only cases where we have not applied an external magnetic field. What happens if an external magnetic field $\mathbf{H}$ is present? The answer is very simple. Instead of (2-15) we just use

$$\mathbf{\Omega}_{\mathbf{H}+\mathring{\mathbf{H}}} = g_0 \frac{e}{2mc} (\mathbf{H} + \mathring{\mathbf{H}}). \qquad (2\text{-}15''')$$

We then find for the angular velocity of precession viewed from the laboratory system instead of (2-18)

$$\mathbf{\Omega}_{\text{lab}} = g_0 \frac{e}{2mc} \mathbf{H} + (g_0 - 1) \frac{e}{2mc} \mathring{\mathbf{H}}. \qquad (2\text{-}18')$$

And for the interaction energy as seen from the laboratory system we obtain instead of (2-19')

$$E_{\mathbf{H}+\text{rel}} = g_0 \frac{e\hbar}{2mc} (\mathbf{H} \cdot \mathbf{s}) + (g_0 - 1) \frac{e\hbar}{2mc} (\mathring{\mathbf{H}} \cdot \mathbf{s}). \qquad (2\text{-}19'')$$

Put into words, (2-19'') states that an electron acts as if it has a magnetic moment $g_0(e\hbar/2mc)$ for an external magnetic field but a moment of $(g_0 - 1)(e\hbar/2mc)$ for the internal magnetic field. Therefore—finally—all the contradictions related to electron spin have been removed.

Thomas published his conclusion as a letter in *Nature* in February 1926 and submitted a full paper to *Philosophical Magazine* in January 1927. This work of Thomas' had a tremendous impact in Europe. After all, the discrepancies between theory and experiment on the level spacing in doublet terms were thereby completely resolved. Also, the fine structures of H, He$^+$, Li^{++}, Be^{+++}

were shown to be special cases of the alkali doublet terms contrary to Sommerfeld's interpretation. Superficially, whereas in Sommerfeld's theory only S, $P, D, \ldots$ are separate for the same n, if we use the alkali model, each of them further splits into two, and therefore it appears as though they have twice as many levels for H, He^+, Li^{++}, Be^{+++} as provided by Sommerfeld's model. This is true. Nevertheless, it was not observed that way, experimentally.

Let me explain the reason for this, taking as an example $n = 2$ for simplicity. The $n = 2$ term of H has an S term, $k = 1$, and a P term, $k = 2$, and the P term in the alkali model is a doublet and therefore is further split into two levels, producing three levels altogether. However, if E is a Coulomb field, the lower component of the doublet level overlaps with the S term.[3] Therefore, there seem to be only two levels. However, the fact that they are overlapping can be confirmed from the intensity of the spectral line. Furthermore, since S and P are overlapping, we observe transitions which would be forbidden by the selection rule if the level were purely S. (It was Goudsmit who first pointed this out.)

For these reasons, H can be discussed side by side with Li and Na, and He with Ne and Ar. (I said before that the K shell was considered to be different from the L shell and the M shell as layers of the core, but behind this line of thought was the idea that H and He are different beasts than Li and Na, and Ne and Ar, respectively.) Furthermore, this is 1927. Heisenberg had already developed matrix mechanics, and Schrödinger had written extensively about the exploits of the equation bearing his name. According to these new theories, a half-integral angular momentum can exist, the magnitude of orbital angular momentum in the atom is not k but $\ell = k - 1$, and the square of the angular momentum is $\ell^2 = \ell(\ell + 1)$ rather than ℓ^2. We can consider a revised *Ersatzmodell* incorporating these new facts along with the idea that multiplicity is caused by a self-rotating electron instead of the core, and we get a model which is closest to Sommerfeld's except that we replace in $(1\text{-}26)_S$ and $(1\text{-}31)_S$

$$j^2 \rightarrow j(j + 1),$$
$$j_a^2 \rightarrow j_a(j_a + 1),$$
$$j_o^2 \rightarrow j_o(j_o + 1).$$

Indeed if we use $(1\text{-}10)$ and express it with Landé's version of the quantum numbers, the peculiar $1/4$ in Landé's g-factor automatically appears. When we add to this model Pauli's exclusion principle, we can develop a very general theory of atomic spectra and the periodic table. Thus all the fog of 1923–24

---

3. These levels are split by the Lamb shift due to the quantum electrodynamical effect, which Tomonaga, Schwinger, and Feynman independently explained. Tomonaga strictly avoids using QED arguments throughout the book.

has completely cleared. One unsolved problem is the tremendous contradiction between a self-rotating electron and classical electron theory. Furthermore, why the angular momentum of self-rotation is $1/2$, why g_0 must be 2, and how Thomas' classical theory can be quantized still remain as problems. Dirac resolved the last three points from a very unexpected direction as I shall explain next time.

Thomas used classical relativity and the classical theory of tops, supplemented by the correspondence principle, to describe the self-rotation of the electron and was able to derive the correct level intervals. After all this even Pauli could not help withdrawing his reservation "classically indescribable" and finally bestowed his sanction upon the postulate of Kronig, Uhlenbeck, and Goudsmit. All's well that ends well!

The relationship among these three persons stemming from their work on the self-rotating electron has many more interesting episodes, but I wish to leave it to another occasion because we do not have time here. I would like to add just one thing: the fact that people later started to call self-rotation *spin*. They say it was Bohr who first used this word. Probably you agree with me that nowadays we do not think of self-rotation or rotation when we hear the word *spin* (except in the case of skating). This is why I tried to avoid saying spin during this talk and purposely said *self-rotation*.

Pauli's Spin Theory and the Dirac Theory

Why Nature Was Not Satisfied by a Simple Point Charge

As I told you last time, an enormous number of ideas, such as the exclusion principle, electron spin, and Thomas' theory, emerged during the period 1925–26. It was also a time of upheaval in physics triggered by the discovery of matrix mechanics by Heisenberg and the development of wave mechanics by Schrödinger. Using these new theories, quantum mechanics was completely rewritten to go beyond the older, incomplete theory. Previously, the correspondence principle had been applied to classical calculations to infer the existence of discrete energy levels and the probability of transitions between them, facts not at all compatible with classical theory.

At the time of discovery, these two new theories looked entirely different from each other in their mathematical formalism as well as in their physical interpretation. Nevertheless, Schrödinger himself soon realized that at least mathematically they are completely equivalent. (Supposedly Pauli also realized this equivalence, but Schrödinger published it earlier.)[1] Toward the end of 1926 Dirac attempted to unify these two theories using a concept called the state vector and established his majestic transformation theory of quantum mechanics.

To explain transformation theory at length would be too much of a digression, so I shall not do that. In essence the idea is to incorporate into an abstract linear space many concepts, such as matrices and their eigenvalues and eigenvectors, which play central roles in matrix mechanics and linear operators and their eigenvalues and eigenvectors, which play central roles in wave mechanics. As you know, a physical quantity is expressed by a matrix in matrix mechanics and by a linear operator in wave mechanics. In this unified theory, on the

1. Eckart independently showed the equivalence of the two theories: Eckart C 1926 Operator Calculus and the Solution of the Equations of Dynamics *Phys. Rev.* **28** 711–726.

Figure 3.1 Paul A. M. Dirac (1902–1984), Leiden 1928. [Photograph by Leningrad Physico-Technical Institute. Courtesy of AIP Emilio Segrè Visual Archives]

other hand, many physical quantities are expressed by abstract linear operators (Dirac's q-numbers). Finding the quantum-mechanically allowed values of a physical quantity is reduced to finding the eigenvalues of the linear operator expressing that physical quantity; in doing this calculation, depending on what kind of orthogonal coordinate is taken in this abstract linear space, the formalism of matrix mechanics or that of wave mechanics emerges. In other words, by transforming orthogonal coordinate systems in the linear space we can derive wave mechanics from matrix mechanics and vice versa. For this reason this all-inclusive theory is called *transformation theory*. Furthermore, in this theory, a state of a mechanical system is expressed by an abstract vector in the linear space; hence this linear space is often called *state space* and the vector the *state vector*.

Much earlier D. Hilbert and later J. von Neumann succeeded in incorporating the mathematics of matrices and vectors and the mathematics of linear operators and functions into that of an abstract linear space, but the number of coordinate axes taken by them in the linear space was either finite or at most countably infinite. Therefore, in their theory we can take only coordinate axes such as the x_1-axis, x_2-axis, x_3-axis, etc., which are indexed by the positive integers. In contrast, Dirac made it possible to use a nondenumerably infinite number of coordinate axes by introducing his well-known δ-function. In other words, in

Dirac's theory we can use an x_q-axis in which the subscript q is a parameter with a continuous range of values. (Actually, mathematicians do not like this type of idea, but it is convenient for physicists.) Therefore, if the axes are countable, the components of a state vector are expressed as

$$\psi_n, \qquad n = 1, 2, 3, \ldots \qquad (3\text{-}1)$$

but if the axes are continuously infinite, the vector can be expressed as

$$\psi(q), \qquad q_1 < q < q_2, \qquad (3\text{-}1')$$

where the function has the variable q as a running variable taking all the values in the domain (q_1, q_2). Of course, we can rewrite (3-1) as

$$\psi(n), \qquad n = 1, 2, 3, \ldots \qquad (3\text{-}1'')$$

and regard the component of the vector as a function where the running variable n takes only discrete numbers, or more generally we can consider the case in which the running variable takes continuous values in a certain domain (q_1, q_2) and then takes discrete values elsewhere.

Now according to the transformation theory, the state vector changes from one instant to another by the transformation called a unitary transformation, but the state vector has to satisfy a first-order differential equation with respect to time. Furthermore, according to Dirac, if you take the principal axes of a certain physical quantity as the coordinate axes in a state space (more specifically, if you take the principal axes of a linear operator expressing that physical quantity), then the state vector is given a definite physical meaning. Depending on whether the physical quantity has discrete or continuous eigenvalues, the principal axes become discrete or continuous, and accordingly the components of the state vector become $\psi(n)$ or $\psi(q)$. In either case, the square of the absolute value of a component of the state vector, that is $|\psi(n)|^2$ (or $|\psi(q)|^2$), gives the probability (or probability density) for the physical quantity to take the nth (or the qth) value. Such is the physical meaning of $\psi(n)$ or $\psi(q)$. (For the continuous spectrum it is convenient to use the eigenvalue of the physical quantity itself as the subscript q. Then the qth value means the value of q.) In this sense, when such coordinate axes are used, each component of the state vector is often called a *probability amplitude*. The usual wave function is the probability amplitude in which the physical quantity determining the coordinate axes in state space is the position of the electrons. In this case the components of the state vector can be written as

$$\psi(\mathbf{x}) = \psi(x, y, z); \qquad (3\text{-}2)$$

then $|\psi(\mathbf{x})|^2$ gives the probability density that the electron is in the vicinity of $\mathbf{x}$. Furthermore, once we know the probability or probability density, we can calculate the expected value of a variety of physical quantities. Suppose there is a physical quantity A with discrete eigenvalues written as A_n for $n = 1, 2, 3, \ldots$ or with continuous eigenvalues in the domain (q_1, q_2) written simply as $q, q_1 < q < q_2$. Let us use as a coordinate system in the state space the principal axes of A, X_n or X_q, and let the component of the state vector with respect to those coordinates be $\psi(n)$ or $\psi(q)$. Then the expectation value of A, $\langle A \rangle$, is given by

$$\langle A \rangle = \sum_n A_n |\psi(n)|^2 + \int_{q_1}^{q_2} q |\psi(q)|^2 dq. \tag{3-3}$$

With the development of these ideas, the framework of quantum mechanics was essentially completed by the end of 1926. There remained the problem of how to incorporate spin and relativity into this grand theoretical framework.

I shall stop here on the general transformation theory so that we can proceed to the story of spin. Heisenberg, who discovered matrix mechanics, immediately applied the new mechanics to the spin model with P. Jordan in 1926. They examined multiplet terms and the anomalous Zeeman effect and obtained correct results on the g-factor, level interval, and fine structure formula. Furthermore, a year later, Pauli applied Dirac's transformation theory, which I just discussed, and attempted to incorporate spin into Schrödinger's representation. Naturally he reached the same conclusion as Heisenberg and Jordan about the problem of multiplet terms. Although this work by Pauli has the same conclusion as Heisenberg's, it has considerable significance in relation to the many-electron problem and the exclusion principle because he used Schrödinger's representation; we cannot disregard its pioneering role for Dirac's subsequent electron theory. Therefore let me talk about Pauli's work for a while.

As you know, the equation discovered by Schrödinger played a central role in wave mechanics. The Schrödinger equation for the mechanical system composed of one electron is

$$\left[H_0 + \frac{\hbar}{i} \frac{\partial}{\partial t} \right] \psi(\mathbf{x}) = 0 \tag{3-4}$$

if we neglect spin, where

$$H_0 = \frac{1}{2m} \mathbf{p}^2 + V(\mathbf{x}), \tag{3-4'}$$

$$\mathbf{p} = (p_x, p_y, p_z) = \left(\frac{\hbar}{i} \frac{\partial}{\partial x}, \frac{\hbar}{i} \frac{\partial}{\partial y}, \frac{\hbar}{i} \frac{\partial}{\partial z} \right),$$

$$\mathbf{x} = (x, y, z),$$

and $\psi(\mathbf{x})$ is the aforementioned probability amplitude, (3-2). This $\psi(\mathbf{x})$ varies with time satisfying (3-4). In order to include spin, what kind of equation should we use instead of (3-4)? This was Pauli's problem.

Since spin is the self-rotation of an electron, one might well at the outset expect that the degree of freedom of self-rotation can be expressed by using some suitable angle φ. In classical mechanics, if the angular velocity of self-rotation takes one definite value, then the only free variable is the direction of the rotation axis. In this case, we can use the azimuthal angle of the self-rotation axis as the canonical coordinate describing the freedom of self-rotation and consider it as our angle φ. If we do this and let $\mathbf{S}$ (the angular momentum expressed by a capital letter should be considered to be in the ordinary units) be the angular momentum for self-rotation, then its z-component S_z will be the canonical momentum conjugate to φ. Therefore, if we take this angle φ as a coordinate for the spin degree of freedom, then it appears that we should use a function $\psi(\mathbf{x}, \varphi)$ in which the angle φ has been incorporated into the wave function instead of $\psi(\mathbf{x})$, and just as the momentum $\mathbf{p}$ conjugate to $\mathbf{x}$ was given by the second formula of (3-4'), we can use as S_z, which is conjugate to φ,[2]

$$S_z = \frac{\hbar}{i} \frac{\partial}{\partial \varphi}. \tag{3-4"}$$

However, this idea introduces one problem.

The difficulty arises from the fact that the magnitude of the spin angular momentum is $(1/2)\hbar$. Namely, this means that the quantum-mechanically allowed values for S_z should be $\pm\hbar/2$, which means that the eigenvalue of S_z is $\pm\hbar/2$ and therefore the eigenfunction of S_z must be $e^{\pm i\varphi/2}$. However, this function does not come back to its original value if φ is varied from 0 to 2π. That is to say, it is a two-valued function, and it is illogical to allow it to be an eigenfunction.

In order to avoid this problem, Pauli proceeded as follows. As I explained before I started talking about spin, in Dirac's transformation theory we can use as the coordinate axes in state space those with discrete subscript n or those with continuous subscript q. Therefore using the angle φ is not the only way to incorporate the spin degree of freedom into the theory. If we write the

---

2. Tomonaga is confused here. If φ is the azimuthal angle of the self-rotation axis, then to be consistent with (3-11), (3-4") cannot be right. This argument is an aside in the text, however, and since it does not affect the text, we leave it here as an example of even a great theoretical physicist making an elementary slip.

wave function as $\psi(\mathbf{x})$, that means we take the principal axes of the position coordinate $\mathbf{x}$ of the electron and that of the spin coordinate φ as the coordinate axes in state space. In order to incorporate the spin degree of freedom, we can use instead of the axis of the angle φ the axis of the self-rotational angular momentum S_z. If we do this, we use as the wave function $\psi(\mathbf{x}, S_z)$. Here S_z is a variable which takes only the two values $\pm\hbar/2$. After all, there is the question of whether we can actually experimentally measure the angle φ anyhow. On the contrary, S_z is directly related to experiment, and for this reason it was very much to Pauli's liking.

From now on we shall use $\mathbf{s}$, measured in units of $\hbar$, rather than the angular momentum $\mathbf{S}$, measured by the usual units. Namely,

$$\mathbf{S} = \hbar\mathbf{s}, \qquad S_x = \hbar s_x, \qquad S_y = \hbar s_y, \qquad S_z = \hbar s_z. \qquad (3\text{-}5)$$

Then our wave function becomes

$$\psi(\mathbf{x}, s_z), \qquad s_z = +\frac{1}{2}, -\frac{1}{2}, \qquad (3\text{-}6)$$

and this function does not have the specter of two-valuedness. The physical meaning of (3-6) is that the probability density of an electron at $\mathbf{x}$ with its spin up is

$$\left| \psi\left(\mathbf{x}, +\frac{1}{2}\right) \right|^2 \qquad (3\text{-}7)_+$$

and with its spin down is

$$\left| \psi\left(\mathbf{x}, -\frac{1}{2}\right) \right|^2. \qquad (3\text{-}7)_-$$

Also, the fact that s_z takes only two values, $\pm 1/2$, means that the state space for the spin degree of freedom is a two-dimensional vector space. If we consider this way, we may say that $\psi(\mathbf{x}, +1/2)$ and $\psi(\mathbf{x}, -1/2)$ are the components of the state vector in the state space.

We now know how to incorporate spin into the wave function $\psi(\mathbf{x})$, but the next problem is how to do the same for Schrödinger's equation. H_0 in (3-4) is the Hamiltonian representing the total energy of the system when spin is neglected. Therefore, in order to incorporate spin into Schrödinger's equation, we should add the energy of spin to H_0. To treat the Zeeman effect, we should add to H_0 the Hamiltonian of interaction between the external magnetic field and the electron.

First let us consider the Hamiltonian involving the external magnetic field **H** and write it H_1. As to the form of H_1, please remember (1-28) and (1-29). We can use as H_1 those expressions which are the interaction energy for the external field and the atom, but we must follow the new quantum mechanics and use $\boldsymbol{\ell}$ instead of **K** and **s** instead of **R**:

$$H_1 = \frac{e\hbar}{2mc} \left[\mathbf{H} \cdot (\boldsymbol{\ell} + g_0 \mathbf{s}) \right]. \qquad (3\text{-}8)$$

Here $\boldsymbol{\ell}$ is the orbital angular momentum in units of Dirac's $\hbar$,

$$\begin{aligned}
\ell_x &= \frac{1}{i} \left(y \frac{\partial}{\partial z} - z \frac{\partial}{\partial y} \right), \\
\ell_y &= \frac{1}{i} \left(z \frac{\partial}{\partial x} - x \frac{\partial}{\partial z} \right), \\
\ell_z &= \frac{1}{i} \left(x \frac{\partial}{\partial y} - y \frac{\partial}{\partial x} \right),
\end{aligned} \qquad (3\text{-}9)$$

and **s** is the spin angular momentum given by (3-5). Furthermore, for spin we have to add the interaction Hamiltonian between the internal magnetic field (2-11') and the spin magnetic moment. If I write that as H_2, then we can write from (2-19')

$$H_2 = (g_0 - 1) \frac{e\hbar}{2mc} (\overset{\circ}{\mathbf{H}} \cdot \mathbf{s}). \qquad (3\text{-}10)$$

Here, the internal magnetic field (2-11') can be written as

$$\overset{\circ}{\mathbf{H}} = \frac{Ze\hbar}{mc} \frac{1}{r^3} \boldsymbol{\ell}, \qquad (3\text{-}10')$$

and therefore

$$H_2 = 2(g_0 - 1)Z \left(\frac{e\hbar}{2mc} \right)^2 \frac{1}{r^3} (\boldsymbol{\ell} \cdot \mathbf{s}). \qquad (3\text{-}10'')$$

The next problem is what type of operator to use for the vector **s**. When we considered H_0 in Schrödinger's equation (3-4), we used as operators for x, y, z simply those coordinates multiplying the wave function from the left, namely $x\cdot, y\cdot, z\cdot$, and for p_x, p_y, p_z we used operators which act on $\psi(x, y, z)$ by $\hbar\partial/i\partial x, \hbar\partial/i\partial y, \hbar\partial/i\partial z$. For ℓ_x, ℓ_y, ℓ_z in H_1 we used the angular momentum operators (3-9). Now the freedom of x, y, z is completely independent of the freedom of spin, and therefore we may consider the same operators acting upon the wave function $\psi(x, y, z, s_z)$ as far as **x**, **p**, and ℓ are concerned.

Now let us consider the spin degree of freedom. As I already told you, according to the general transformation theory, the independent variable q, which is used in the wave function $\psi(q)$, is an eigenvalue of some physical quantity (for example, the variables x, y, z used in $\psi(x, y, z)$ are the eigenvalues of the physical quantity called the position coordinate of the particle), and the operator which expresses that physical quantity is simply the operator q, which multiplies the wave function $\psi(q)$ by q. For the physical quantity p, which is conjugate to q, we use $\hbar\partial/i\partial q$ (assuming that the conjugate quantity exists). Therefore for the wave function $\psi(x, y, z, s_z)$, we can use an operator for the z-component of the spin angular momentum $s_z\cdot$. Note that the Hamiltonian operators H_1 and H_2 contain s_x and s_y in addition to the physical quantity s_z. What operators should these be which are applied to ψ? There is the following way of doing it. (This was not the way Pauli adopted.)

We considered the angle φ as a coordinate which describes the spin degree of freedom. When we used this as one of the arguments in the wave function, the function became two-valued, and therefore we switched to s_z, which is conjugate to φ, as the running variable of the wave function. Now if we use this φ, then we can express s_x and s_y by

$$s_x = \sqrt{s^2 - s_z^2}\,\cos\varphi \qquad s_y = \sqrt{s^2 - s_z^2}\,\sin\varphi, \qquad (3\text{-}11)$$

where φ and $\hbar s_z$ are conjugate to each other. Are the operators we are looking for not those in (3-11) when φ is replaced by $\partial/i\partial s_z$? We then have the problem of how to define $\cos(\partial/i\partial s_z)$ and $\sin(\partial/i\partial s_z)$. It is not impossible to define these, but since it is too complicated, Pauli did not adopt this method. Instead he considered as follows.

In matrix mechanics the angular momentum matrices m_x, m_y, m_z satisfy in general the commutation relations

$$\begin{aligned} m_x m_y - m_y m_x &= i m_z, \\ m_y m_z - m_z m_y &= i m_x, \\ m_z m_x - m_x m_z &= i m_y. \end{aligned} \qquad (3\text{-}12)$$

Furthermore, $|\mathbf{m}|^2 = m_x^2 + m_y^2 + m_z^2$ has the eigenvalue

$$|\mathbf{m}|^2 = m(m+1), \quad m = 0, 1, 2, \ldots \quad \text{or } m = \frac{1}{2}, \frac{3}{2}, \frac{5}{2}, \ldots \qquad (3\text{-}13)$$

Therefore Pauli also requires equations similar to (3-12) for the spin angular momentum (s_x, s_y, s_z),

$$s_x s_y - s_y s_x = i s_z,$$
$$s_y s_z - s_z s_y = i s_x, \qquad \qquad (3\text{-}12')$$
$$s_z s_x - s_x s_z = i s_y,$$

and assumes as a special characteristic of spin, instead of (3-13),

$$|\mathbf{s}|^2 = \frac{1}{2}\left(\frac{1}{2}+1\right) = \frac{3}{4}. \qquad \qquad (3\text{-}13')$$

Therefore he introduced the 2×2 matrices which are called "Pauli's matrices"

$$\sigma_x = \begin{pmatrix} 0 & 1 \\ 1 & 0 \end{pmatrix}, \quad \sigma_y = \begin{pmatrix} 0 & -i \\ i & 0 \end{pmatrix}, \quad \sigma_z = \begin{pmatrix} 1 & 0 \\ 0 & -1 \end{pmatrix}, \qquad (3\text{-}14)$$

and set

$$s_x = \frac{1}{2}\sigma_x, \qquad s_y = \frac{1}{2}\sigma_y, \qquad s_z = \frac{1}{2}\sigma_z. \qquad (3\text{-}14')$$

Then we can easily find that these s_x, s_y, s_z satisfy (3-12') as well as (3-13'). Therefore Pauli proposed to use the matrices (3-14') as the matrices of s_x, s_y, s_z in (3-8) and (3-10). Let us note that Pauli's matrices have the properties

$$\sigma_\mu^2 = 1 \qquad \qquad (\mu = x, y, z), \qquad \qquad (3\text{-}14'')$$
$$\sigma_\mu \sigma_\nu + \sigma_\nu \sigma_\mu = 0 \qquad (\mu \neq \nu; \mu, \nu = x, y, z).$$

Now let me explain Pauli's suggestion more concretely. I told you earlier that the wave function is in the form of (3-6), where, we note, the variable s_z takes only two values, $+1/2$ and $-1/2$. We can then consider the functions $\psi(\mathbf{x}, +1/2)$ and $\psi(\mathbf{x}, -1/2)$ to be components of a two-component quantity, or we can consider that the wave function is a function which has two components. We then can write

$$\psi = \begin{pmatrix} \psi\left(\mathbf{x}, +\frac{1}{2}\right) \\ \psi\left(\mathbf{x}, -\frac{1}{2}\right) \end{pmatrix}. \qquad \qquad (3\text{-}15)$$

This way of writing (3-15) corresponds to considering the state vector (which is a two-dimensional vector) in the spin state space to be a column vector. And applying s_x, s_y, s_z (or $\sigma_x, \sigma_y, \sigma_z$) to the wave function ψ is nothing but multiplying the matrices (3-14') or (3-14) to the column vector. This was Pauli's proposal. According to this idea,

$$\sigma_x \psi = \begin{pmatrix} 0 & 1 \\ 1 & 0 \end{pmatrix} \begin{pmatrix} \psi\left(\frac{1}{2}\right) \\ \psi\left(-\frac{1}{2}\right) \end{pmatrix}$$

$$\sigma_y \psi = \begin{pmatrix} 0 & -i \\ i & 0 \end{pmatrix} \begin{pmatrix} \psi\left(\frac{1}{2}\right) \\ \psi\left(-\frac{1}{2}\right) \end{pmatrix} \tag{3-16}$$

$$\sigma_z \psi = \begin{pmatrix} 1 & 0 \\ 0 & -1 \end{pmatrix} \begin{pmatrix} \psi\left(\frac{1}{2}\right) \\ \psi\left(-\frac{1}{2}\right) \end{pmatrix},$$

and therefore

$$s_x \psi = \begin{pmatrix} \frac{1}{2}\psi\left(-\frac{1}{2}\right) \\ \frac{1}{2}\psi\left(\frac{1}{2}\right) \end{pmatrix}$$

$$s_y \psi = \begin{pmatrix} -\frac{i}{2}\psi\left(-\frac{1}{2}\right) \\ \frac{i}{2}\psi\left(\frac{1}{2}\right) \end{pmatrix} \tag{3-16'}$$

$$s_z \psi = \begin{pmatrix} \frac{1}{2}\psi\left(\frac{1}{2}\right) \\ -\frac{1}{2}\psi\left(-\frac{1}{2}\right) \end{pmatrix}.$$

Here s_z applied to ψ is indeed $\psi(s_z)$ multiplied by s_z. (For this discussion the variable $\mathbf{x}$ is irrelevant, and I stopped writing it explicitly.)

This is Pauli's idea. If we consider it this way, the Schrödinger equation

$$\left[H_0 + H_1 + H_2 + \frac{\hbar}{i} \frac{\partial}{\partial t} \right] \psi(\mathbf{x}, s_z) = 0 \tag{3-17}$$

is a set of simultaneous differential equations for the two functions $\psi(\mathbf{x}, +1/2)$ and $\psi(\mathbf{x}, -1/2)$ and hereafter will be called the Pauli equation. For a stationary state we replace ψ with

$$\psi = \phi e^{-iEt/\hbar}$$

and determine the energy of the state by solving

$$[H_0 + H_1 + H_2 - E] \phi(\mathbf{x}, s_z) = 0. \tag{3-17'}$$

By this procedure we can calculate the level intervals within the doublet term and the anomalous Zeeman effect. We have used as H_0 the nonrelativistic Hamiltonian, and therefore the fine structure formula cannot be derived correctly, but we can make the approximation a little better on this point. Pauli himself points out clearly that his theory is intrinsically nonrelativistic since the spin degree of freedom is expressed as s_x, s_y, s_z, which is a vector in x, y, z space only. He pointed out that in order to make his theory relativistic, he had to introduce the antisymmetric tensor (a six-element vector) in Minkowski space, but judging from Thomas' theory, the electron has only a magnetic moment in its rest system, and therefore half of the six elements must be zero when the electron is at rest. Pauli abandoned trying to create a relativistic theory, saying that it is extremely difficult to apply such a condition to the spin degree of freedom. Furthermore, just as in the theory of Heisenberg and Jordan, Pauli's theory introduces the electron spin angular momentum of $1/2$ and g_0 factor of 2 into H_1 and H_2 arbitrarily, and the Thomas factor $1/2$ is also introduced ad hoc into H_2. For these reasons Pauli knew his theory was tentative.

In his paper Pauli examined how the two components of (3-15) are transformed by rotation around the x, y, and z axes based on the invariance of his equation (3-17) with respect to rotation, and he noticed that the transformation is remarkably different from that of a vector. Pauli's discussion eventually leads to the double-valued representation of the rotation group and spinors, but I shall discuss these topics later.

Pauli also considers many-electron systems, but here we limit ourselves to deriving an important theorem from Pauli's theory which is related to my next lecture. The theorem is:

> When there are two electrons (as long as the particle has spin $1/2$, it could be any particle), the wave function for which the sum of the two spins equals 1 does not change its value when the spin variables of the electrons are exchanged (that is, the function is symmetric). The wave function for which the sum becomes 0 changes sign when the spin variables are interchanged (namely, the function is antisymmetric).

The converse of this theorem is also valid.

Proof: Let us take the spin matrices of two electrons as $\mathbf{s}_1 = (s_{1x}, s_{1y}, s_{1z})$ and $\mathbf{s}_2 = (s_{2x}, s_{2y}, s_{2z})$. Then the square of the magnitude of the total spin $\mathbf{s}_1 + \mathbf{s}_2$ equals

$$\left|\mathbf{s}_1 + \mathbf{s}_2\right|^2 = \left|\mathbf{s}_1\right|^2 + \left|\mathbf{s}_2\right|^2 + 2(\mathbf{s}_1 \cdot \mathbf{s}_2),$$

and if we use (3-13'),

$$= \frac{1}{2}\left[3 + 4(\mathbf{s}_1 \cdot \mathbf{s}_2)\right]. \tag{I}$$

Now we write the wave function $\psi(s_{1z}, s_{2z})$ in the form of a column vector

$$\psi = \begin{pmatrix} \psi\left(\ \frac{1}{2},\ \ \frac{1}{2}\right) \\ \psi\left(\ \frac{1}{2}, -\frac{1}{2}\right) \\ \psi\left(-\frac{1}{2},\ \ \frac{1}{2}\right) \\ \psi\left(-\frac{1}{2}, -\frac{1}{2}\right) \end{pmatrix} \qquad \text{(II)}$$

(omitting the variables x_1 and x_2 because they are not related to this problem) and multiply the matrices $s_1 \cdot s_2 = s_{1x}s_{2x} + s_{1y}s_{2y} + s_{1z}s_{2z}$. We then get, using (3-16'),

$$(s_1 \cdot s_2)\psi = \begin{pmatrix} \frac{1}{4}\psi\left(\ \frac{1}{2},\ \ \frac{1}{2}\right) \\ \frac{1}{2}\psi\left(-\frac{1}{2},\ \ \frac{1}{2}\right) - \frac{1}{4}\psi\left(\ \frac{1}{2}, -\frac{1}{2}\right) \\ \frac{1}{2}\psi\left(\ \frac{1}{2}, -\frac{1}{2}\right) - \frac{1}{4}\psi\left(-\frac{1}{2},\ \ \frac{1}{2}\right) \\ \frac{1}{4}\psi\left(-\frac{1}{2}, -\frac{1}{2}\right) \end{pmatrix}.$$

If we use equation (I) here, we get

$$|s_1 + s_2|^2\psi = \begin{pmatrix} 2\psi\left(\ \frac{1}{2},\ \ \frac{1}{2}\right) \\ \psi\left(\ \frac{1}{2}, -\frac{1}{2}\right) + \psi\left(-\frac{1}{2},\ \ \frac{1}{2}\right) \\ \psi\left(-\frac{1}{2},\ \ \frac{1}{2}\right) + \psi\left(\ \frac{1}{2}, -\frac{1}{2}\right) \\ 2\psi\left(-\frac{1}{2}, -\frac{1}{2}\right) \end{pmatrix}. \qquad \text{(III)}$$

Now the total spin equals 1 means that from (3-13') $|s_1 + s_2|^2\psi = 1(1+1)\psi = 2\psi$, and the total spin equals 0 means $|s_1 + s_2|^2\psi = 0$. In the former case, if we compare (II) and (III), we find that it is necessary and sufficient if

$$\psi\left(\frac{1}{2}, -\frac{1}{2}\right) + \psi\left(-\frac{1}{2}, \frac{1}{2}\right) = 2\psi\left(\frac{1}{2}, -\frac{1}{2}\right).$$

This means

$$\psi\left(\frac{1}{2}, -\frac{1}{2}\right) = \psi\left(-\frac{1}{2}, \frac{1}{2}\right)$$

and therefore

$$\psi(s_{1z}, s_{2z}) = \psi(s_{2z}, s_{1z}). \qquad \text{(IV)}_+$$

This equation is a necessary and sufficient condition for the total spin to be 1. In the latter case, it is necessary and sufficient that

$$\psi\left(\frac{1}{2}, \frac{1}{2}\right) = \psi\left(-\frac{1}{2}, -\frac{1}{2}\right) = 0,$$

$$\psi\left(\frac{1}{2}, -\frac{1}{2}\right) + \psi\left(-\frac{1}{2}, \frac{1}{2}\right) = 0.$$

Thus the condition we need is

$$\psi(s_{1z}, s_{2z}) = -\psi(s_{2z}, s_{1z}) \tag{IV}_-$$

Q. E. D.

I talked at length about Pauli's theory, so let us proceed now. As I told you earlier, Pauli himself was never satisfied with this theory. Perhaps he racked his brains trying to make his theory relativistic, but probably for him it looked too difficult. For these reasons the perfectionist had no other choice than to publish a paper in which the spin was introduced ad hoc. It was Dirac, however, who in one stroke solved these two problems of introducing spin and of deriving a relativistic equation.

The problem of making quantum mechanics relativistic had already been touched upon by Schrödinger in 1926 when he published the fourth paper in the famous series of papers "Quantization as Eigenvalue Problems I, II, III, IV." The same type of equation he considered was also discussed almost simultaneously by O. Klein and W. Gordon. All these ideas are to propose the so-called Klein-Gordon equation in paper IV by Schrödinger instead of the nonrelativistic equation in paper I.

They proposed this because the equation is the simplest one for a free particle (this means a particle without an external field) which yields the de Broglie–Einstein relation

$$\nu^2 - \left(\frac{c}{\lambda}\right)^2 = \left(\frac{mc^2}{h}\right)^2, \quad E = h\nu, \quad p = \frac{h}{\lambda}. \tag{3-18}$$

Here ν is the frequency of the de Broglie wave and λ its wavelength. Schrödinger found that he could not derive a result consistent with Sommerfeld's fine structure formulas, contrary to expectation, when he tried to use this equation to determine the hydrogen energy levels. Referring to this point in the fourth paper, he mentions that this discrepancy would be removed if spin were taken into account, but he did not proceed further. If I may digress here, de Broglie discussed the idea of the phase wave along the line of relativity, and he discovered the relation (3-18) as a result. For this reason, they say Schrödinger initially wanted to create wave mechanics from the Klein-Gordon equation, which appears in his fourth paper. However, because he could not get the hydrogen energy levels consistent with experiment, he postponed addressing relativity and treated first the nonrelativistic hydrogen atom in his first paper.

Now let us discuss Dirac's achievement. As you know, the Klein-Gordon equation is

$$\left[\left(-\frac{\hbar}{i}\frac{\partial}{c\partial t} + \frac{e}{c}A_0\right)^2 - \sum_{r=1}^{3}\left(\frac{\hbar}{i}\frac{\partial}{\partial x_r} + \frac{e}{c}A_r\right)^2 - m^2c^2\right] \quad (3\text{-}19)$$
$$\times \psi(x, y, z, t) = 0.$$

Here A_0 and A_r are the components of the four-vector potential of the externally applied electromagnetic field. If as a special case we set $A_0 = A_r = 0$, then we obtain the formula without an external field, and the equation has a solution of a plane wave

$$\psi(x, y, z, t) = e^{2\pi i(z/\lambda - \nu t)}.$$

If we substitute this into (3-19), we find that the relation

$$\left(\frac{h\nu}{c}\right)^2 - \left(\frac{h}{\lambda}\right)^2 = m^2c^2$$

must be satisfied. This is nothing but the de Broglie–Einstein equation (3-18). For this reason it was thought that (3-19) is a fundamental equation of relativistic wave mechanics.

Dirac, however, questioned this idea. First of all, this equation is a differential equation second-order with respect to time. The wave function in wave mechanics should have the meaning of probability amplitude, and according to his transformation theory, it must satisfy a differential equation first-order with respect to time. (Indeed, both the Schrödinger equation (3-4) and the Pauli equation (3-17) are differential equations first-order with respect to time.)

Dirac therefore required the wave equation to be a differential equation first-order with respect to time even when relativity is taken into account. Now in relativity, the time variable t and the space variables x, y, z must be treated equivalently, and therefore the equation sought should also be first-order with respect to the space variables. If this is the case, the formula should not contain the square of $(-\hbar\partial/ic\partial t + e/c\,A_0)$ and $(\hbar\partial/i\partial x_r + e/c\,A_r)$ like (3-19), and therefore it must be of the form

$$\left[\left(-\frac{\hbar}{i}\frac{\partial}{c\partial t}+\frac{e}{c}A_0\right)-\sum_{r=1}^{3}\alpha_r\left(\frac{\hbar}{i}\frac{\partial}{\partial x_r}+\frac{e}{c}A_r\right)-\alpha_0 mc\right] \quad (3\text{-}20)$$
$$\times\,\psi(x,y,z,t)=0.$$

Now we have these α_r and α_0 in the formula. How shall we determine them? Here Dirac used his extraordinary genius.

He thought that for a free particle without the external field A_0 and A_r, the de Broglie–Einstein relation (3-18) must be satisfied and therefore ψ must be a solution of the Klein-Gordon equation. First in order to derive a second-order equation from (3-20) for a free particle, that is,

$$\left[-\frac{\hbar}{i}\frac{\partial}{c\partial t}-\sum_{r=1}^{3}\alpha_r\frac{\hbar}{i}\frac{\partial}{\partial x_r}-\alpha_0 mc\right]\psi(x,y,z,t)=0, \quad (3\text{-}21)$$

we apply to it an operator

$$\left[-\frac{\hbar}{i}\frac{\partial}{c\partial t}+\sum_{r=1}^{3}\alpha_r\frac{\hbar}{i}\frac{\partial}{\partial x_r}+\alpha_0 mc\right]. \quad (3\text{-}21)$$

We obtain

$$\left[-\frac{\hbar}{i}\frac{\partial}{c\partial t}+\sum_{r=1}^{3}\alpha_r\frac{\hbar}{i}\frac{\partial}{\partial x_r}+\alpha_0 mc\right]$$
$$\cdot\left[-\frac{\hbar}{i}\frac{\partial}{c\partial t}-\sum_{r=1}^{3}\alpha_r\frac{\hbar}{i}\frac{\partial}{\partial x_r}-\alpha_0 mc\right]\psi(x,y,z,t)=0. \quad (3\text{-}21')$$

Dirac demanded that this equation be the Klein-Gordon equation. Such a requirement cannot be fulfilled if α_0, α_1, α_2, α_3 are ordinary numbers. However, Dirac found that it *is* possible to satisfy this requirement if they are matrices. In this case, (3-21') becomes

$$\left[\left(-\frac{\hbar}{i}\frac{\partial}{c\partial t}\right)^2-\sum_{r=1}^{3}\alpha_r^2\left(\frac{\hbar}{i}\frac{\partial}{\partial x_r}\right)^2\right.$$
$$\left.-\sum_{\mu<\nu}(\alpha_\mu\alpha_\nu+\alpha_\nu\alpha_\mu)\left(\frac{\hbar}{i}\right)^2\frac{\partial^2}{\partial x_\mu\partial x_\nu}-\alpha_0^2 m^2 c^2\right]\psi=0, \quad (3\text{-}21'')$$

and, therefore, if the matrices α_0, α_1, α_2, α_3 have the properties

$$\alpha_\mu^2 = 1 \qquad\qquad (\mu = 0, 1, 2, 3) \qquad (3\text{-}22)$$

$$\alpha_\mu\alpha_\nu + \alpha_\nu\alpha_\mu = 0 \qquad (\mu \neq \nu; \mu, \nu = 0, 1, 2, 3),$$

then the requirement is fulfilled. He then introduced

$$\alpha_0 = \begin{pmatrix} 1 & 0 & 0 & 0 \\ 0 & 1 & 0 & 0 \\ 0 & 0 & -1 & 0 \\ 0 & 0 & 0 & -1 \end{pmatrix} \alpha_1 = \begin{pmatrix} 0 & 0 & 0 & 1 \\ 0 & 0 & 1 & 0 \\ 0 & 1 & 0 & 0 \\ 1 & 0 & 0 & 0 \end{pmatrix}$$

$$\alpha_2 = \begin{pmatrix} 0 & 0 & 0 & -i \\ 0 & 0 & i & 0 \\ 0 & -i & 0 & 0 \\ i & 0 & 0 & 0 \end{pmatrix} \alpha_3 = \begin{pmatrix} 0 & 0 & 1 & 0 \\ 0 & 0 & 0 & -1 \\ 1 & 0 & 0 & 0 \\ 0 & -1 & 0 & 0 \end{pmatrix} \quad (3\text{-}23)$$

as the simplest matrices that satisfy (3-22) and used them in (3-20).

When these 4×4 matrices are introduced into (3-20), ψ accordingly becomes a column vector with four components

$$\psi = \begin{pmatrix} \psi_1 \\ \psi_2 \\ \psi_3 \\ \psi_4 \end{pmatrix}, \qquad (3\text{-}24)$$

and therefore (3-20) becomes a set of simultaneous differential equations with respect to the four functions $\psi_1, \psi_2, \psi_3, \psi_4$. It is interesting to compare the matrices (3-23) and Pauli's matrices (3-14); we immediately notice that Dirac's 4×4 matrices may be written using Pauli's 2×2 matrices (3-14), the 2×2 unit matrix, and the zero matrix

$$\mathbf{1} = \begin{pmatrix} 1 & 0 \\ 0 & 1 \end{pmatrix} \qquad \mathbf{0} = \begin{pmatrix} 0 & 0 \\ 0 & 0 \end{pmatrix} \qquad (3\text{-}25)$$

as

$$\alpha_0 = \begin{pmatrix} \mathbf{1} & \mathbf{0} \\ \mathbf{0} & -\mathbf{1} \end{pmatrix} \qquad \alpha_1 = \begin{pmatrix} \mathbf{0} & \sigma_1 \\ \sigma_1 & \mathbf{0} \end{pmatrix}$$

$$\alpha_2 = \begin{pmatrix} \mathbf{0} & \sigma_2 \\ \sigma_2 & \mathbf{0} \end{pmatrix} \qquad \alpha_3 = \begin{pmatrix} \mathbf{0} & \sigma_3 \\ \sigma_3 & \mathbf{0} \end{pmatrix} \qquad (3\text{-}23')$$

(Here we use the subscripts 1, 2, 3 instead of the subscripts x, y, z.) Accordingly, it is convenient to write the four-component column vector (3-24) as

$$\psi = \begin{pmatrix} \psi^+ \\ \psi^- \end{pmatrix} \tag{3-24'}$$

using the two-component column vectors

$$\psi^+ = \begin{pmatrix} \psi_1 \\ \psi_2 \end{pmatrix} \qquad \psi^- = \begin{pmatrix} \psi_3 \\ \psi_4 \end{pmatrix}. \tag{3-26}$$

Dirac applied his equation to the case of a central force and showed that the correct (that is, with the correct Thomas factor) energy level spacing can be derived in the first-order approximation. Moreover, he showed that the orbital angular momentum ℓ is not a conserved quantity and that a conserved quantity is obtained only when $\frac{1}{2} \begin{pmatrix} \sigma & 0 \\ 0 & \sigma \end{pmatrix}$ is added to it. Thus a conserved quantity cannot be obtained from the orbital angular momentum alone but is obtained only when $\frac{1}{2} \begin{pmatrix} \sigma & 0 \\ 0 & \sigma \end{pmatrix}$ is added, meaning that the electron has spin angular momentum and that

$$\text{spin angular momentum} \ = \frac{1}{2} \begin{pmatrix} \sigma & 0 \\ 0 & \sigma \end{pmatrix} \quad \text{(in units of } \hbar \text{)}. \tag{3-27}$$

Moreover, Dirac found that if an external field is present, then, unlike the case for a free particle, the procedure used for obtaining (3-21') from (3-21) does not yield the Klein-Gordon equation. Dirac found that this difference has the form of the interaction between the external field and the spin magnetic moment. In this argument, the expression $-\begin{pmatrix} \sigma & 0 \\ 0 & \sigma \end{pmatrix}$ appears in the part of the formula expected for the spin magnetic moment, and therefore

$$\text{spin angular momentum} \ = -\begin{pmatrix} \sigma & 0 \\ 0 & \sigma \end{pmatrix} \quad \text{(in units of } e\hbar/2mc \text{)}, \tag{3-27'}$$

and this means $g_0 = 2$. Dirac further discussed how the four-component quantity (3-24) should be transformed in order for his equation to be Lorentz invariant. This is a generalization of Pauli's idea about the two-component quantity. Thus Dirac started only from the requirements of relativity and transformation theory and, without using any ad hoc assumption, showed that the spin angular momentum of the electron, its magnetic moment, and the Thomas factor can all be derived correctly.

We mortals are left reeling by this staggering outpouring of ideas from Dirac. His first requirement is that the equation be linear in time. Such a thought had not occurred to Schrödinger, Klein, Gordon, or even Pauli. Even if it had been realized, it is beyond our humble expectations that this requirement lead to the electron spin and to the determination of its magnetic moment. They say Pauli often referred to Dirac's thought process as acrobatic, and the development of the famous equation bearing Dirac's name shows this characteristic most clearly. He seems to have been euphoric about this work as can be sensed from the following sentences in the introduction of his paper.

Namely, he starts the paper saying that Pauli and C. G. Darwin had already discussed electron spin. He says, "The question remains as to why Nature should have chosen this particular model [such as the spinning electron] for the electron instead of being satisfied with the point-charge." He continues, "It appears that the simplest Hamiltonian for a point-charge electron satisfying the requirements of both relativity and the general transformation theory leads to an explanation of all duplexity phenomena without further assumption."

However, on another occasion Dirac also mentioned the following (this was in his lecture in 1969 when he was awarded the Oppenheimer Memorial Prize). He calculated the problem of the alkali atom only to the first order of approximation in this paper. He did not try to carry out an accurate calculation for the hydrogen atom and derive the fine structure formula which was then known to agree with experiment perfectly. He did not do that because he was afraid that if he calculated that much and happened to obtain a result that did not agree with the formula, he would thus find that his theory was not correct. Because of his immense confidence in the correctness of his theory, even a 1 percent fear that his expectation might be crushed would perhaps also have been very big. This sentiment is as illogical as that of a sick person who does not want to consult a doctor for fear of being told he has cancer. However, even theoretical physicists often become prisoners of such irrationally twisted sentiment. By the way, the work to derive the fine structure formula was done by Darwin and Gordon.

This work of Dirac's is definitely the work of genius; however, I believe it was very much stimulated by Pauli's previous work. Dirac's idea to use the matrix which satisfies (3-22) in order to relate the first-order equation (3-20) and the Klein-Gordon equation may have been catalyzed by the fact that Pauli's matrices satisfy (3-14"). Also, the adoption of the idea of the simultaneous differential equations using the four-component quantity (3-24) may have been hinted at by Pauli's creation of the simultaneous differential equations (3-17) using the two-component quantity (3-15). In fact, Dirac points out explicitly in his paper that Pauli's $\sigma_r (r = 1, 2, 3)$ satisfy (3-22). He further states that if we take Pauli's 2×2 matrices, we can find only three which satisfy (3-22), and therefore he considers 4×4 matrices. Furthermore, Dirac's way of proving the Lorentz invariance of (3-20) is precisely the generalization of Pauli's proof that Pauli's equation is invariant with respect to rotation around the x, y, z axes. For these reasons I said earlier that Pauli's work is the forerunner of Dirac's. Dirac's work was completed in January 1928, less than one year after Pauli's work.

If this work of Dirac's clarified the reason why Nature was not satisfied by a simple point charge but required a charge with spin, it might emerge as a natural conclusion that the proton will also have a spin $\hbar/2$ and a magnetic moment $e\hbar/2m_p c$ (m_p is the mass of the proton.) In the actual history, however, the fact that the proton has spin $\hbar/2$ had already been discovered half a year or so before Dirac's work was published. Now from a layman's point of view, the establishment of the proton spin emerged from a very unexpected problem. That was the problem of the specific heat of hydrogen at low temperature. There is an interesting episode on how proton spin was established from specific heat, but let me talk about it in the next lecture.

Finally, let me add some comments, which you might or might not find as superfluous as legs on a snake. Just as Pauli's matrices σ_x, σ_y, σ_z were related to the x, y, z components of spin angular momentum, we can regard the six matrices

$$\frac{1}{i}\alpha_0\alpha_2\alpha_3 \qquad \frac{1}{i}\alpha_0\alpha_3\alpha_1 \qquad \frac{1}{i}\alpha_0\alpha_1\alpha_2 \qquad (3\text{-}28)$$
$$i\alpha_0\alpha_1 \qquad i\alpha_0\alpha_2 \qquad i\alpha_0\alpha_3$$

formed from Dirac's α-matrices as components of one six-vector, and we can regard them as the relativistic generalization of electron spin. In fact, if we use (3-23), the first three are

$$\frac{1}{i}\alpha_0\alpha_2\alpha_3 = \begin{pmatrix} \sigma_1 & 0 \\ 0 & -\sigma_1 \end{pmatrix} \qquad \frac{1}{i}\alpha_0\alpha_3\alpha_1 = \begin{pmatrix} \sigma_2 & 0 \\ 0 & -\sigma_2 \end{pmatrix} \qquad (3\text{-}28')$$
$$\frac{1}{i}\alpha_0\alpha_1\alpha_2 = \begin{pmatrix} \sigma_3 & 0 \\ 0 & -\sigma_3 \end{pmatrix},$$

and they look apparently different from the spin angular momentum (3-27), but their expectation values for an electron at rest agree with those of (3-27). Moreover, we can prove that the three quantities

$$i\alpha_0\alpha_1 = \begin{pmatrix} 0 & i\sigma_1 \\ -i\sigma_1 & 0 \end{pmatrix} \qquad i\alpha_0\alpha_2 = \begin{pmatrix} 0 & i\sigma_2 \\ -i\sigma_2 & 0 \end{pmatrix} \qquad (3\text{-}28'')$$
$$i\alpha_0\alpha_3 = \begin{pmatrix} 0 & i\sigma_3 \\ -i\sigma_3 & 0 \end{pmatrix},$$

have expectation values of 0 for an electron at rest. I told you before that Pauli realized the necessity to relativistically six-dimensionalize his spin matrices but had great difficulty introducing the condition that half of them must be 0 in the rest system. Dirac, by using 4×4 matrices, has done this quite simply. Thus Dirac has derived everything about electron spin through Lorentz invariance and that the wave equation must be first order without using a model at all.

It may be since this work of Dirac's that we started not to think about self-rotation or rotation from the words *electron spin*. (It is a different matter for nuclear spin.) In any case, if the real nature of electron spin is something like this, it is truly "classically indescribable," is it not?

Proton Spin

Three Scholars Tackle the Hydrogen Molecule

Now how did the specific heat of the hydrogen molecule and proton spin come to be connected?

We can say that right from the beginning, quantum mechanics had an unseverable relation to the problem of specific heat. For example, it may be said that the problem associated with blackbody radiation was how to eliminate the infinite value of the "specific heat of vacuum." Also, the fact that the specific heat of a diatomic molecule is $(3/2)R$ at low temperature and not $(5/2)R$ as predicted from classical theory was first explained only by quantum mechanics. This is because the rotational energy of molecules is quantized, and therefore if $kT/2$, which should be distributed according to the equipartition law, becomes smaller than the quantum of rotation, then rotational motion cannot occur.

A diatomic molecule is a mechanical system in which electrons are moving around two nuclei that are arranged like a dumbbell. Therefore, its energy is given by the sum of four energies: the electronic energy E_{el}, the vibrational energy E_{vib} corresponding to the expansion and contraction of the distance between the nuclei, the overall molecular rotational energy E_{rot}, and the translational energy of the center of gravity E_{tr}.

$$E = E_{el} + E_{vib} + E_{rot} + E_{tr} \qquad (4\text{-}1)$$

Here we have neglected the interaction between different motions, but this has little effect on the following discussion.

Of these four energies only E_{tr} and E_{rot} contribute to the low-temperature specific heat. (Since both E_{el} and E_{vib} have large quanta, we consider sufficiently low temperatures such that $kT/2$ is much smaller than these quanta.) Now since E_{tr} is not quantized, its contribution to the specific heat is $(3/2)R$ all the way down to 0 K. In other words, if we consider the specific heat of one molecule instead of a mole of molecules, then the contribution from translational motion is $(3/2)k$. Therefore, only E_{rot} is in question.

Now please remember the elementary quantum mechanics of rotational energy. According to it, we obtain

$$E_{rot} = \frac{\hbar^2}{2I} J(J+1), \qquad J = 0, 1, 2, \ldots \qquad (4\text{-}2)$$

for the rotational energy, where I is the moment of inertia of the dumbbell. Here J is the rotational quantum number, which gives the angular momentum measured in units of $\hbar$. For a given J, the state is degenerate with a multiplicity

$$g(J) = 2J + 1 \qquad (4\text{-}3)$$

corresponding to the $2J+1$ directions of quantization of the angular momentum.

Now except for hydrogen nuclei, the nuclei forming the dumbbell are particles with structure. Then, as Pauli once considered, nuclei may have spin or may not be spherical. (Around 1924, Pauli assumed that atomic nuclei have spin to explain the hyperfine structure of spectral lines.) If that is the case, then when the two nuclei are connected in a dumbbell fashion, we have to consider in which direction the nuclei are oriented with respect to the dumbbell axis. We then must consider, in addition to the direction of the dumbbell itself with respect to space, a fourth degree of freedom corresponding to the direction of each nucleus with respect to the dumbbell axis. Namely, just as we gave the electron the spin degree of freedom, we also have to give nuclei a fourth degree of freedom related to their orientation. (We do not know yet whether this is the direction of spin or, if the nuclei are nonspherical, the direction of the ellipsoids.) Then there is an energy related to the fourth degree of freedom (written, say, E_4), and we must use instead of (4-1)

$$E = E_{el} + E_{vib} + E_{rot} + E_{tr} + E_4; \qquad (4\text{-}1')$$

the contribution of E_4 will naturally appear in the specific heat, and therefore by examining the specific heat we can get information on this new degree of freedom.

This proposition sounds fine, but actually only the first half is correct; the second half is not good as stated. The reason is that, even if such a degree of freedom existed, its energy is extremely small and does not affect the specific heat. If this energy were large enough to affect the specific heat, its effect should naturally appear in the band spectra of molecules, but this is not experimentally observed. (The fourth degree of freedom of the *electron* clearly appears as a doubling of the spectrum. But even that does not affect the specific heat.) The connection between the specific heat and the fourth degree of freedom results from an entirely different physical effect.

What is this effect? It is a discovery as important in the development of quantum mechanics as that of electron spin; that is, the discovery of the relation between the statistics of a particle and the symmetry of its wave function. As you know, there are two kinds of particles that appear in quantum mechanics: those obeying Bose statistics (bosons) and those obeying Fermi statistics (fermions). Around 1927, in the period I am now discussing, only the photon and the electron were clearly known to belong to such groups; the former is representative of bosons and the latter of fermions. There was a conjecture that the α-particle is very likely a boson. I might note here that saying a fermion obeys Fermi statistics is equivalent to saying the particle follows Pauli's exclusion principle.

In the third lecture I said that two problems remained when Dirac's transformation theory was completed. One was spin, and the other was making the theory relativistic, but now I find I forgot to mention another one—namely, how to incorporate into quantum mechanics these two types of statistics.

Once again, Dirac and Heisenberg came to the rescue and answered this problem in 1926. How the two types of statistics were discovered and how they were incorporated into quantum mechanics is a very interesting subject in itself, but let me cut it short and give only the conclusion. Dirac and Heisenberg found that in a mechanical system composed of identical particles such as electrons or photons, if the wave function is symmetric with respect to an interchange of particles and only such states are realized, then the assembly of particles obeys Bose statistics, and if the wave function is antisymmetric with respect to an interchange of particles and only such states are realized, then the assembly of particles obeys Fermi statistics. This discovery played a very important role in problems of mechanical systems composed of many identical particles, namely, in many-body problems. That story is also extremely interesting, but I will touch on one such example in the next lecture. Now let us return to molecules.

The diatomic molecule we are considering is a mechanical system composed of two nuclei and several electrons (in the case of H_2, two electrons). In the approximation that the total energy is expressed by (4-1'), its wave function can be expressed as the product of the wave functions of each motion

$$\psi = \psi_{\text{el}} \cdot \psi_{\text{vib}} \cdot \psi_{\text{rot}} \cdot \psi_{\text{tr}} \cdot \psi_4 = \phi \psi_4. \qquad (4\text{-}4)$$

Here ϕ is the product of the wave functions ψ_{el}, ψ_{vib}, ψ_{rot}, and ψ_{tr}, which correspond to already known degrees of freedom. The wave function ψ_4 corresponds to the fourth degree of freedom, whose identity remains a mystery at this stage. It is a function of the two coordinates that describe the fourth degree of freedom for the two nuclei.

Now in the elementary theory of rotating molecules, which I gave you earlier, the state of a molecule is expressed just by ψ_{rot} and ψ_{tr} in (4-4). Sometimes it is sufficient to discuss the problem of low-temperature specific heat by this simple theory. (If there is a fourth degree of freedom, we add ψ_4.) However, for discussing band spectra, ψ_{el} also plays an important role, and a more advanced theory is necessary. These advanced theories of diatomic molecules have been developed by various people, but Hund particularly made a big contribution to the treatment of ψ_{el} and the theoretical interpretation of experimental results. Hund also triggered the quest to obtain information on the fourth degree of freedom of the proton from the specific heat of the hydrogen molecule. Since a substantial effort is required to go into the detailed molecular theory here, let us simply proceed with the elementary theory mentioned earlier.

Looking at Hund's study, we find that if the state of the electron is $^1\Sigma$ according to his notation (you may think of this as similar to the ^1S state of atoms), then the elementary theory ignoring ψ_{el} can be applied to our problem. For example, in the $^1\Sigma$ state, E_{rot} is given by (4-2), and the degeneracy of the level is given by (4-3). (We ignore the fourth degree of freedom for the time being.) Furthermore, just as in the elementary theory, ψ_{rot} can be expressed as

$$\psi_{\text{rot}} = P_J^M(\cos\Theta)e^{iM\Phi} \quad M = J, J-1, J-2, \ldots, -J+1, -J, \quad (4\text{-}5)$$

using the angles Θ and Φ to specify the orientation of the dumbbell axis.

Here we limit our scope to homonuclear molecules such as H_2 or O_2. The reason is that in those cases, the statistical problem of nuclei (whether they are bosons or fermions) which I referred to earlier plays an important role, and which statistics is followed affects the rotational specific heat of the molecule through the existence of the fourth degree of freedom. Specifically, if the two nuclei are bosons, ψ must be symmetric with respect to the interchange of the nuclei, and if they are fermions, ψ must be antisymmetric. This is reflected in the specific heat for the following reasons.

Let us consider the factor ϕ in ψ in (4-4) and examine whether ϕ is symmetric or antisymmetric with respect to the interchange of two nuclei. Since only the spatial coordinates of the two nuclei are involved in ϕ and the enigmatic fourth coordinate is not included, we can examine the symmetry of ϕ from the known theory. Especially when the electron is in the $^1\Sigma$ state, we can use elementary theory and easily determine the symmetry of ϕ. Luckily, in H_2 the ground state of the electron is $^1\Sigma$. (By contrast, in O_2 it is $^3\Sigma$.)

Let us take the spatial coordinates of the two nuclei as x_1 and x_2. In the elementary theory, we ignore ψ_{el}, leaving

$$\phi = \psi_{\text{tr}} \cdot \psi_{\text{vib}} \cdot \psi_{\text{rot}}, \quad (4\text{-}6)$$

in which,

$$\psi_{tr} \text{ is a function of } \frac{\mathbf{x}_1 + \mathbf{x}_2}{2}$$
$$\psi_{vib} \text{ is a function of } |\mathbf{x}_1 - \mathbf{x}_2| \qquad (4\text{-}7)$$
$$\psi_{rot} \text{ is a function of } \frac{\mathbf{x}_1 - \mathbf{x}_2}{|\mathbf{x}_1 - \mathbf{x}_2|}.$$

It is easily seen that $(\mathbf{x}_1 + \mathbf{x}_2)/2$ and $|\mathbf{x}_1 - \mathbf{x}_2|$ are invariant with respect to interchange of $\mathbf{x}_1$ and $\mathbf{x}_2$. Hence, ψ_{tr} and ψ_{vib} are always symmetric with respect to the interchange of $\mathbf{x}_1$ and $\mathbf{x}_2$. $(\mathbf{x}_1 - \mathbf{x}_2)/|\mathbf{x}_1 - \mathbf{x}_2|$, in contrast, changes sign when $\mathbf{x}_1$ and $\mathbf{x}_2$ are interchanged. Therefore, it is not that obvious whether ψ_{rot} is symmetric or antisymmetric. The following argument shows that it is symmetric if J is even and antisymmetric if J is odd.

In order to see this, we introduce angles Θ and Φ to represent the orientation of the unit vector $(\mathbf{x}_1 - \mathbf{x}_2)/|\mathbf{x}_1 - \mathbf{x}_2|$. Namely, let the spherical polar coordinates of $(\mathbf{x}_1 - \mathbf{x}_2)/|\mathbf{x}_1 - \mathbf{x}_2|$ be $(1, \Theta, \Phi)$. Now, suppose we interchange $\mathbf{x}_1$ and $\mathbf{x}_2$ on the left-hand side of (4-5), which is

$$\psi_{rot}\left(\frac{\mathbf{x}_1 - \mathbf{x}_2}{|\mathbf{x}_1 - \mathbf{x}_2|}\right) = P_J^M(\cos\Theta)e^{iM\Phi}.$$

Since $(\mathbf{x}_2 - \mathbf{x}_1)/|\mathbf{x}_2 - \mathbf{x}_1| = -(\mathbf{x}_1 - \mathbf{x}_2)/|\mathbf{x}_1 - \mathbf{x}_2|$, if the polar coordinates of $(\mathbf{x}_1 - \mathbf{x}_2)/|\mathbf{x}_1 - \mathbf{x}_2|$ are $(1, \Theta, \Phi)$, the coordinates of $-(\mathbf{x}_1 - \mathbf{x}_2)/|\mathbf{x}_1 - \mathbf{x}_2|$ are obviously $(1, \pi - \Theta, \pi + \Phi)$, and therefore we have

$$\psi_{rot}\left(\frac{\mathbf{x}_2 - \mathbf{x}_1}{|\mathbf{x}_2 - \mathbf{x}_1|}\right) = P_J^M[\cos(\pi - \Theta)]e^{iM(\pi+\Phi)}.$$

On the other hand, as is well known,

$$P_J^M[\cos(\pi - \Theta)]e^{iM(\pi+\Phi)} = P_J^M(-\cos\Theta)(-1)^M e^{iM\Phi}$$
$$= (-1)^J P_J^M(\cos\Theta)e^{iM\Phi},$$

and thus we have

$$\psi_{rot}\left(\frac{\mathbf{x}_2 - \mathbf{x}_1}{|\mathbf{x}_2 - \mathbf{x}_1|}\right) = (-1)^J \psi_{rot}\left(\frac{\mathbf{x}_1 - \mathbf{x}_2}{|\mathbf{x}_1 - \mathbf{x}_2|}\right).$$

This means that ψ_{rot} is symmetric with respect to interchange of $\mathbf{x}_1$ and $\mathbf{x}_2$ if J is even and antisymmetric if J is odd.

In summary, we can conclude that, for the function ϕ which is a product of ψ_{tr}, ψ_{vib}, and ψ_{rot},

$$\phi \text{ is} \begin{cases} \text{symmetric with respect to interchange of } \mathbf{x}_1 \text{ and } \mathbf{x}_2 \text{ if } J \text{ is even} \\ \text{antisymmetric with respect to interchange of } \mathbf{x}_1 \text{ and } \mathbf{x}_2 \text{ if } J \text{ is odd} \end{cases} \quad (4\text{-}8)$$

It might be easy to understand these conclusions if they are depicted as in figure 4.1. A of this figure shows the rotational energy levels given by (4-2). The solid lines show levels which are symmetric with respect to the nuclear permutation, and the dashed lines are levels which are antisymmetric. I shall discuss the meaning of B later.

It took some time to prepare for our discussion, but now we may proceed to the problem of whether the nucleus is a boson or a fermion. Let us first assume that the nucleus does not have a spin and furthermore that it is spherically symmetric. For this case we do not need the fourth degree of freedom. Therefore, $\psi = \phi$, and the symmetry of ϕ is at the same time the symmetry of ψ. Since from the conclusion of (4-8) only levels with symmetric ψ exist if the nucleus is a boson, the dashed levels of figure 4.1A do not exist. If the nucleus is a fermion, then only levels with antisymmetric ψ appear, and thus the solid-line levels of figure 4.1A do not exist. For these reasons the levels which are realized are those given in figure 4.2.

If this is the case, the rotational specific heat will naturally be different depending on whether all the levels in figure 4.1A appear, only the left-hand

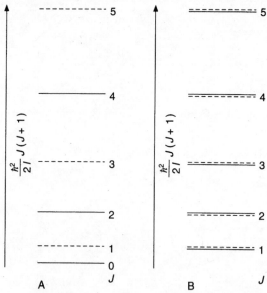

Figure 4.1. (A) Rotational levels for $^1\Sigma$. (B) Rotational levels for $^1\Pi$.

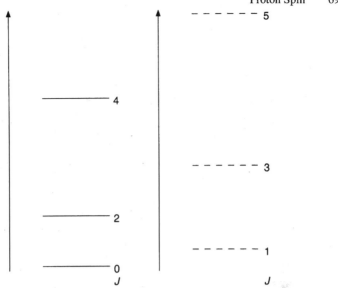

Figure 4.2 Rotational levels for bosons (left) and rotational levels for fermions (right).

levels in figure 4.2 appear, or only the right-hand levels of figure 4.2 appear. Also, in a band spectrum, if the levels are as shown on either the left or right side of figure 4.2, then quite a few levels drop off. Therefore, by measuring the specific heat or analyzing the band, we can immediately determine whether the nucleus is a boson or a fermion.

However, the situation is a bit more complicated if there is a fourth degree of freedom, for example, spin. In this case, $\psi \neq \phi$, and we cannot use (4-8) directly to discuss the symmetry of ψ. In this case $\psi = \phi \cdot \psi_4$, and since (symmetric function) $\times$ (symmetric function) = (symmetric function), (symmetric function) $\times$ (antisymmetric function) = (antisymmetric function), and (antisymmetric function) $\times$ (antisymmetric function) = (symmetric function), it is obvious that

if ϕ is symmetric, then in order for $\begin{cases} \psi \text{ to be symmetric, } \psi_4 \text{ must be symmetric} \\ \psi \text{ to be antisymmetric, } \psi_4 \text{ must be antisymmetric} \end{cases}$

if ϕ is antisymmetric, then in order for $\begin{cases} \psi \text{ to be symmetric, } \psi_4 \text{ must be antisymmetric} \\ \psi \text{ to be antisymmetric, } \psi_4 \text{ must be symmetric.} \end{cases}$

Then we combine this with (4-8) and conclude

if nucleus is a boson $\begin{cases} \text{——— level is paired with a symmetric } \psi_4 \\ \text{- - - - - level is paired with an antisymmetric } \psi_4 \end{cases}$ (4-9)

$$\text{if nucleus is a fermion} \begin{cases} \text{——— level is paired with an antisymmetric } \psi_4 \\ \text{- - - - - level is paired with a symmetric } \psi_4 . \end{cases}$$

For these reasons, if the nucleus has the fourth degree of freedom, then neither solid nor dashed levels disappear, but in each, ψ_4 has a different symmetry. Depending on whether ψ_4 is symmetric or antisymmetric, the energy E_4 may differ slightly, but the difference is extremely small, and we can consider (4-2) to hold as it is. Therefore, the position of the levels is just as in figure 4.1A.

However, it is quite possible that the degeneracy of level E_4 may differ depending on whether ψ_4 is symmetric or antisymmetric. For such cases the degeneracy of E as a whole in (4-1') is different, and this difference might be detected by experiment. For example, the intensity of spectral lines will be stronger if the degeneracy of a level is larger. Also, if we calculate the specific heat using statistical mechanics, this degeneracy becomes the statistical weight and thus affects the result. It was this link that Hund spotted.

The case of spin provides a good example of the degeneracy of level E_4 depending on the symmetry of ψ_4. For example, suppose the spins of the two nuclei are each $1/2$. According to the theorem which I proved in the third lecture, if ψ_4 is symmetric with respect to interchange of the spin coordinates, the total spin is 1, and therefore triple degeneracy will occur related to the number of azimuthal quantizations of spin. If ψ_4 is antisymmetric with respect to the interchange of spin coordinates, the total spin is 0, and there is no degeneracy (namely, the degeneracy is 1). These are the hard facts.

Now in general we write the degeneracy of level E_4 as

$$g_s \text{ if } \psi_4 \text{ is symmetric and}$$
$$g_a \text{ if } \psi_4 \text{ is antisymmetric.}$$

Then from (4-9) we conclude that the degeneracy of level E_4 is

$$\begin{cases} \text{if the nucleus is a boson, then} \begin{cases} g_s \text{ for a ——— level} \\ g_a \text{ for a - - - level} \end{cases} \\ \text{if the nucleus is a fermion, then} \begin{cases} g_a \text{ for a ——— level} \\ g_s \text{ for a - - - level} \end{cases} \end{cases} \quad (4\text{-}10)$$

So far we assumed that the electronic state is $^1\Sigma$ and derived (4-10) using an elementary theory. However, we see from Hund's theory that the conclusion of (4-10) holds for any electronic state. The only difference is that the rotational-level scheme differs from that of figure 4.1A. I give you only one example, shown in figure 4.1B. This is the case in which the electronic state is not $^1\Sigma$ but the state which is called $^1\Pi$ in Hund's notation. (This is similar to ^1P in atomic states.) As you can see from the figure, the energies of the levels are also given by (4-2) even in the case of a $^1\Pi$ state, but the $J = 0$ level does not exist, and $J = 1, 2, 3, \ldots$ Also, each rotational level is already doubly degenerate (in addition to the degeneracy of $2J + 1$ from the azimuthal quantization) even before considering the fourth degree of freedom, and the components are depicted as one solid and one dashed. In figure 4.1B I express this degeneracy by $\equiv$, using a solid line and a dashed line drawn very close together.

So regardless of whether the electronic state is $^1\Sigma$ or $^1\Pi$, if the degeneracies of the solid and dashed levels are different, then that difference will appear in the intensities of band spectral lines. Specifically, if the nucleus is a boson, the ratio of the intensity of the line corresponding to a transition from a solid level to a solid level to that from a dashed level to a dashed level should be $g_s : g_a$, and if the nucleus is a fermion, it should be $g_a : g_s$. The transition from a solid level to a dashed level is forbidden in our approximation. (In a higher-order approximation, the transition becomes allowed, but with an extremely small probability.) If we reiterate this conclusion using J and supposing that the terminating level is $^1\Sigma$, then transitions to levels with $J = 0, 1, 2, 3, 4, \ldots$ will have the intensity alternation $g_s : g_a : g_s : g_a : \ldots$ if the nucleus is a boson and an alternation of $g_a : g_s : g_a : g_s : \ldots$ if the nucleus is a fermion. Therefore, for example, if the fourth degree of freedom of the nucleus is spin and if it is $1/2$, then from the last lecture

$$g_s = 3 \qquad \text{and} \qquad g_a = 1, \qquad (4\text{-}11)$$

and therefore the intensity alternation will be

Figure 4.3 Friedrich Hund (1896–), 1927. [Courtesy of AIP Emilio Segrè Visual Archives, Franck Collection]

$$\text{for a boson} \quad 3 : 1 : 3 : 1 : \dots \qquad (4\text{-}11')$$

$$\text{for a fermion} \quad 1 : 3 : 1 : 3 : \dots$$

For these reasons, if we examine the band spectrum of the hydrogen molecule for such an intensity alternation and determine the intensity ratio, then we can determine the spin of the proton and also whether protons are bosons or fermions. When Hund developed this idea in early 1927, there were unfortunately not enough data on the H_2 band spectrum. Therefore, he decided to use the already-known experimental value of the rotational specific heat of the hydrogen molecule.

Let us consider hydrogen gas at a temperature T. According to statistical mechanics, the average rotational energy for one molecule is given as

$$\langle E_{\text{rot}} \rangle = \frac{1}{Z} \left[\sum_{J=\text{even}} \beta g(J) E(J) e^{\frac{-E(J)}{kT}} + \sum_{J=\text{odd}} g(J) E(J) e^{\frac{-E(J)}{kT}} \right] \quad (4\text{-}12)$$

$$Z = \sum_{J=\text{even}} \beta g(J) e^{\frac{-E(J)}{kT}} + \sum_{J=\text{odd}} g(J) e^{\frac{-E(J)}{kT}}.$$

Here $E(J)$ and $g(J)$, given by (4-2) and (4-3), respectively, are the rotational energy and the degeneracy of azimuthal quantization, and

$$\beta = \frac{\text{degeneracy of } E_4 \text{ when } J \text{ even}}{\text{degeneracy of } E_4 \text{ when } J \text{ odd}}. \quad (4\text{-}13)$$

If we remember (4-10),

$$\beta = \frac{g_s}{g_a} \quad \text{if the proton is a boson} \quad (4\text{-}13')$$

$$= \frac{g_a}{g_s} \quad \text{if the proton is a fermion.}$$

So if the fourth degree of freedom of the proton is spin and its value is $1/2$, then we should use (4-11) for g_s and g_a. Therefore

$$\beta = 3 \quad \text{if the proton is a boson} \quad (4\text{-}13'')$$

$$= \frac{1}{3} \quad \text{if the proton is a fermion,}$$

but for the time being, β is undetermined.

When Hund was calculating the specific heat, the moment of inertia I of the hydrogen molecule was also unknown. Therefore he gave β and I a variety of values, calculated $\langle E_{rot} \rangle$ by (4-12), and determined the specific heat by the formula

$$C_{rot} = \frac{d\langle E_{rot} \rangle}{dT}. \tag{4-14}$$

From this he found that

$$\beta \approx 2 \text{ and } I \approx 1.54 \times 10^{-41} \text{g cm}^2 \tag{4-15}$$

gave the best fit to the experimental C_{rot} at low temperature. Supposing that the spin of the proton is $1/2$, this value of β does not agree with either the value 3 (assuming the proton to be a boson) or the value $1/3$ (assuming the proton to be a fermion). He said in his paper that the agreement with experiment is not that bad for $\beta = 3$ if $I = 1.69 \times 10^{-41}$g cm^2, but he refrained from concluding that the proton is a boson with spin $1/2$. Hund finished this work in Bohr's institute in Copenhagen and published it in February 1927.

While Hund was whipping this up, a physicist by the name of Hori arrived in Copenhagen. This scholar was an experimental spectroscopist, and in short he is Takeo Hori, who graduated from Kyoto University and became a professor in Hokkaido University.[1] He was then at the Lu Shun Institute of Technology, and he went to Copenhagen to make the most of his expertise in spectroscopy. We can imagine that the problem of the hydrogen molecule was a hot topic there because Hund was also in Copenhagen at that time. So Bohr suggested to Hori that he study the band spectrum of H_2. Hori therefore performed a very elaborate experiment and a laborious analysis on the band spectrum of the hydrogen molecule which had been discovered by Werner only in the previous year, 1926. As a result, he discovered the intensity alternation in the hydrogen rotational spectrum. He found that the intensities of spectral lines are

weaker for transitions between levels with even J and
stronger for transitions between levels with odd J. $\qquad$ (4-16)

He did not state the intensity ratio, but if we judge from the figure given by him, it looks like 1:3. As I shall tell you later, when a scholar by the name of D. M. Dennison asked him about the intensity ratio, Hori answered that it was approximately 1:3.

---

1. Tomonaga could have said much more here since he was personally close to Hori. Hori was one of Tomonaga's physics instructors, and he also married Tomonaga's older sister.

Figure 4.4 Japanese studying in Bohr's institute, 1926, from the left: Yoshio Nishina, Shin-ichi Aoyama (1882–1959), Takeo Hori (1899–1994), and Kenjiro Kimura (1896 –1988).

This result does not agree at all with the conclusion that Hund obtained from the specific heat. From the specific heat Hund obtained $\beta \approx 2$, which means that spectral lines with even J have to be stronger than those with odd J by about a factor of 2. This is clearly opposite to Hori's result (4-16). Furthermore, Hori could determine the moment of inertia from the band spectrum and obtained

$$I = 4.67 \times 10^{-41} \text{g cm}^2. \tag{4-16'}$$

This is also very different from Hund's result (4-15). Hori pointed these out.

When Hori was summarizing his experimental results at Copenhagen, Dennison of the University of Michigan was aroused by the discrepancy between Hund's and Hori's results. Therefore, he went to Copenhagen and asked to read the manuscript of Hori's paper prior to publication. As a result, he got an idea.

Dennison noticed that Hori had not found any line corresponding to the transition from solid to dashed levels. As I told you earlier, this transition is forbidden in the first approximation, but it may be weakly allowed in a higher approximation. If this is the case, there could be a weak spectral line corresponding to the transition, but in fact there is none. This means that the probability of a transition from solid to dashed levels is extremely small. Furthermore, Dennison also realized that theoretically this probability must be very small. If this is the case, he thought, then a long time is required for a hydrogen molecule to undergo the transition from an even J state to an odd J state and vice versa, so much so that the time for the experimental measurement of specific heat is too short for the transition to occur during the experiment. In other words, the experiment may be conducted without establishing thermal equilibrium between molecules with even J and those with odd J. If this is the case, it is natural to expect that Hund's conclusion does not agree with experiment because the formula (4-12) which Hund used was based on the assumption of thermal equilibrium. This was Dennison's idea.

Therefore, he calculated the specific heat by considering the gas with even J and the gas with odd J as separate gases which do not interconvert. According to this idea, we can write the rotational energy of the gas with even J and that of the gas with odd J as $E_{\text{rot}}^{\text{even}}$ and $E_{\text{rot}}^{\text{odd}}$, respectively, and calculate $\langle E_{\text{rot}}^{\text{even}} \rangle$ and $\langle E_{\text{rot}}^{\text{odd}} \rangle$ separately. They are

$$\langle E_{\text{rot}}^{\text{even}} \rangle = \frac{1}{Z_{\text{even}}} \sum_{J=\text{even}} g(J)E(J)\, e^{\frac{-E(J)}{kT}}$$

$$Z_{\text{even}} = \sum_{J=\text{even}} g(J)\, e^{\frac{-E(J)}{kT}}; \qquad (4\text{-}17)_{\text{even}}$$

$$\langle E_{\text{rot}}^{\text{odd}} \rangle = \frac{1}{Z_{\text{odd}}} \sum_{J=\text{odd}} g(J)E(J)\, e^{\frac{-E(J)}{kT}}$$

$$Z_{\text{odd}} = \sum_{J=\text{odd}} g(J)\, e^{\frac{-E(J)}{kT}}. \qquad (4\text{-}17)_{\text{odd}}$$

If we make the mixing ratio of the two gases $\rho : 1$, then the energy of the gas mixture is obviously

$$\langle E_{\text{rot}} \rangle = \frac{\rho}{\rho+1} \langle E_{\text{rot}}^{\text{even}} \rangle + \frac{1}{\rho+1} \langle E_{\text{rot}}^{\text{odd}} \rangle. \qquad (4\text{-}18)$$

Therefore, we calculate the specific heat $C_{\text{rot}} = d\langle E_{\text{rot}} \rangle / dT$ by using this $\langle E_{\text{rot}} \rangle$.

Based on this idea, Dennison calculated C_{rot} for a variety of values of ρ and I, trying to find the values that give the best computed C_{rot}. He thus found

$$\rho = \frac{1}{3} \quad \text{and} \quad I = 4.64 \times 10^{-41} \text{g cm}^2. \qquad (4\text{-}19)$$

The value of $\rho = 1/3$ agrees with the spectral intensity ratio he learned from Hori, and also the value of I, 4.64, agrees very well with Hori's 4.67.

Dennison concludes his paper by stating that the discrepancy between Hund's calculation and Hori's experiment had been completely resolved and left it at that.

However, two weeks after he submitted the paper, that is June 3, 1927, he sent an addendum to the paper dated June 16, 1927. In this addendum he pointed out for the first time that the mixing ratio of $1/3 : 1$ means that the spin of the proton is $1/2$. He first says that this ratio of $1/3 : 1$ is the mixing ratio of J-even and J-odd gas at room temperature and concludes that this means precisely that the proton is a fermion of spin $1/2$.

I am not absolutely sure, but I suspect from these points that Dennison initially did not realize that his work was a very important one related to the proton spin. Indeed, he chose as the title of the paper "A Note on the Specific Heat of the Hydrogen Molecule," a rather modest title, and took a style as if he were simply commenting on the papers of Hund and Hori. I suspect that after he had submitted the paper, he realized the importance of his work and added the note added June 16. . . . [2]

So I would also make an addendum here, added . . . It is whether the ρ of $\rho : 1$ agrees with β in (4-13'). The reason I worry about this is that only when they agree can we conclude from Dennison's $\rho = 1/3$ that $\beta = 1/3$ and from that can we say using (4-13") that the proton is a fermion with spin $1/2$. The argument goes as follows.

As you probably know, according to statistical mechanics, if hydrogen gas is in thermal equilibrium,

$$\rho(T) : 1 = \sum_{J=\text{even}} \beta g(J) \, e^{\frac{-E(J)}{kT}} : \sum_{J=\text{odd}} g(J) \, e^{\frac{-E(J)}{kT}}.$$

In order for this to agree with $\beta : 1$, we must have

$$\sum_{J=\text{even}} g(J) \, e^{\frac{-E(J)}{kT}} = \sum_{J=\text{odd}} g(J) \, e^{\frac{-E(J)}{kT}},$$

and this is not in general true. However, if T is sufficiently large so that $kT \gg \hbar^2/2I$, this relation holds to a sufficiently good approximation. For this reason,

$$\rho(\text{room } T) = \beta.$$

As Dennison says, if his ρ is the mixing ratio at room temperature, then we can conclude from $\rho = 1/3$ that the proton is a fermion with spin $1/2$. Then

---

2. According to Dennison's account, he knew full well its import: Dennison D M 1974 Recollections of Physics and Physicists during the 1920's *Am. J. Phys.* **42** 1051–1056.

on what ground did Dennison interpret ρ not as $\rho(T)$ but as $\rho(\text{room } T)$? It is precisely at this point that Dennison differed from Hund.

According to Dennison's idea, the hydrogen molecules used for the experiment had been kept at room temperature for a long time. Therefore, before the experiment was started, the mixing ratio was $\rho(\text{room } T) : 1$. During the experiment, this gas was put into a flask, and the specific heat was measured by cooling the gas to a very low temperature. However, even after cooling in the flask, the mixing ratio remains at $\rho(\text{room } T) : 1$ and not $\rho(T) : 1$ because the transition between solid and dashed levels takes a very long time. This is Dennison's key idea. If instead $\rho(T)$ is used for ρ, then $\langle E_{\text{rot}} \rangle$ of (4-18) becomes the same as that of (4-12) used by Hund. (4-12) is by no means incorrect, but Hund forgot to examine whether the condition needed for that form was actually satisfied by the experiment and therefore got the wrong answer. Even Hund made such a mistake![3] You should be very careful in using formulas.

I have talked long enough, so I conclude here the episode of Hund, Hori, and Dennison tackling the hydrogen molecule. Perhaps I spoke too long, but I thought most of the audience did not know how the proton spin and statistics were determined. Subsequently, O. Stern determined not only the spin of the proton but also its magnetic moment by a direct method (bending a molecular beam with a magnetic field, his specialty). That was 1933, several years later.

3. Tomonaga is being unfair to write that "Hund forgot to examine . . ." It was beyond anyone's imagination at that time that the ortho- and para-hydrogen require such a long time to interconvert. It was the separation of ortho- and para-H_2 by Bonhoeffer and Harteck [Bonhoeffer K F and Harteck P 1929 Über Para- und Orthowasserstoff Z. Phys. Chem. **B4** 113–141] and by Eucken and Hiller [Eucken A and Hiller K 1929 Der Nachweis einer Umwandlung durch Messung der spezifischen Wärme Z. Phys. Chem. **B4** 142–168] which gave evidence for Dennison's argument.

Interaction Between Spins

From the Helium Spectrum to Ferromagnetism

In lecture 4 I discussed the hydrogen molecule. Compared with the great mainstream of the development of quantum mechanics, which I talked about in lectures 1 through 3, the research on molecules is like a small eddy near a riverbank. However, the spectacle of leaves whirling in such an eddy is also quite captivating. As I told you in the previous lecture, it often happens that sometimes from this recess in a riverbank a current such as proton spin $= 1/2$ comes out and joins the main stream.

However, it also often happens that a current near the bank makes its own channel, and more water collects in the area until the channel grows to be a large, separate river that branches out from the main stream. As you know, from around 1927 to 1928 after the framework of quantum mechanics had been established, solid state physics rapidly grew like a branch of such a river and developed into a gigantic new discipline.

Today I would like to reflect on how to treat the interaction between electron spins when a mechanical system has many electrons. If I do that, I shall necessarily touch upon a major problem in solid state physics—the problem of ferromagnetism, which had been awaiting an answer for a long time.

The spin of the electron was discovered from the problem of atomic spectra. I discussed this in lectures 1 and 2, taking as examples the simplest alkali metal and alkaline earth. Now for the case of an alkali atom, there is only one electron outside the closed shell, and therefore from the study of its spectrum, we cannot find the key to solving a problem specific to a system composed of two or more electrons. Several phenomena have been observed in such multiple-electron systems which lead one to think that the interaction between electrons is very strong, and this has long been a puzzle. Since the electron has a magnetic moment associated with its spin, two electrons will naturally interact with each other through the magnetic interaction, but experiments show that the spin interaction is apparently four or five orders of magnitude larger than one might expect.

The simplest example of this is seen in the spectra of He and alkaline earths. As I told you in lecture 1, the spectral terms of alkaline earths may be classified into singlet terms and triplet terms, and I repeat in figure 5.2 the levels of Na as representative of alkali metals and the levels of Mg as representative of alkaline earths. You should refer to the figure in this lecture.

As I told you in lecture 2, the levels of an alkaline atom are doubled (the S term is an exception) because the spin of the electrons outside the closed shells is quantized along two directions in the atom. In other words, if we are in the electron rest system, the nucleus rotates around the electron and thus produces a magnetic field at the position of the electron (I called this magnetic field in lecture 2 the internal magnetic field $\mathbf{\overset{\circ}{H}}$), and therefore the electron spins are quantized as $\pm 1/2$ with respect to this internal magnetic field, and the levels split into two. In lecture 1 we considered this splitting using the old model of the core, and in lecture 2 we abandoned the idea of core and the idea of electron spin emerged, but I did not talk about the actual relation between the quantum number j and spin. With the advent of the new quantum mechanics, the

Figure 5.1 Werner K. Heisenberg (1901–1976). [Courtesy of AIP Emilio Segrè Visual Archives]

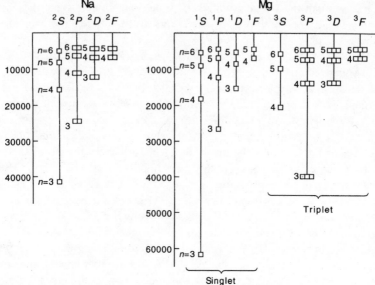

Figure 5.2 Spectral terms for alkalis (left) and alkaline earths (right).

usage of quantum numbers has changed a little bit, so let me continue talking, supplementing this point.

First, the shape and size of the orbit are determined by n and ℓ, and corresponding to these the orbital energy of the electron is given by

$$E_\ell = E^{(1)}(n, \ell). \tag{5-1}$$

According to the new quantum mechanics, the magnitude of the orbital angular momentum is ℓ in units of $\hbar$. Therefore, we will write the angular momentum itself as $\boldsymbol{\ell}$. Next we consider the electron spin. The spin angular momentum is $\mathbf{s}$, and its magnitude s is $1/2$. The vector $\mathbf{s}$ is quantized parallel or antiparallel to the internal magnetic field. Now since the direction of the internal magnetic field coincides with the direction of $\boldsymbol{\ell}$, this conclusion may be rephrased: the spin vector $\mathbf{s}$ is quantized either along the direction of $\boldsymbol{\ell}$ or opposite to the direction of $\boldsymbol{\ell}$.

Therefore, if I write the total angular momentum of the atom as $\mathbf{j}$, then it is the sum of $\boldsymbol{\ell}$ and $\mathbf{s}$

$$\mathbf{j} = \boldsymbol{\ell} + \mathbf{s}, \tag{5-2}$$

and the values which j can take are

$$j = \begin{cases} \ell + \frac{1}{2} & \text{if } \mathbf{s} \text{ is oriented along } \boldsymbol{\ell} \\ \ell - \frac{1}{2} & \text{if } \mathbf{s} \text{ is oriented opposite to the direction of } \boldsymbol{\ell}. \end{cases} \tag{5-3}$$

We assume here that

$$\ell \neq 0. \tag{5-3'}$$

If $\ell = 0$, obviously

$$j = \frac{1}{2}. \tag{5-3"}$$

For $\ell \neq 0$ the vector $\mathbf{s}$ is quantized along $\boldsymbol{\ell}$ or opposite to $\boldsymbol{\ell}$ as shown in figure 5.3.

Now what happens to the energy of the atom? We have in addition to the energy of orbital motion (5-1) the energy of interaction between the internal magnetic field and magnetic moment. Let us write this energy $E_{s,\ell}$ because it is the energy due to the interaction between the orbital motion and the spin. This depends on the shape of the orbit and the direction of $\mathbf{s}$ with respect to $\boldsymbol{\ell}$ and therefore depends on $\mathbf{j}$: it is a function of n, ℓ, and j

$$E_{s,\ell} = E_{s,\ell}(n, \ell; j). \tag{5-4}$$

Therefore, the total energy of the atom is

$$E_{\text{tot}}(n, \ell, j) = E^{(1)}(n, \ell) + E_{s,\ell}(n, \ell; j). \tag{5-5}$$

If $\ell \neq 0$, j takes the two values of (5-3), and E_{tot} becomes a doublet term; if $\ell = 0$, j is only $1/2$, so E_{tot} is singlet. This is a supplement to lecture 1. Please

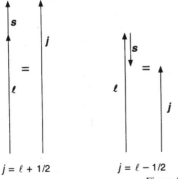

$j = \ell + 1/2$ $j = \ell - 1/2$

Figure 5.3 Doublet terms due to $\mathbf{s} + \boldsymbol{\ell}$.

look at the Na levels in figure 5.2, and you will see this is the case. The real form of $E_{s,\ell}$ is given by the expectation value $\langle H_2 \rangle$ of (3-10") in lecture 3, but I will not write it here because a qualitative discussion suffices.

Next, what happens in the alkaline earths? As seen in the Mg levels of figure 5.2 and as I told you in lecture 1, the levels can be classified into two groups; in one group all levels are singlet, and in the other group all are triplet except for $\ell = 0$. In lecture 1, we named these two classifications of levels singlet terms and triplet terms.

Now can we understand this phenomenon using $\ell + s$ as we just did for alkali metals? About this problem, the following was the accepted theory until 1926.

In alkaline earth atoms, there are two electrons outside the closed shell. In the Mg levels shown in figure 5.2, one electron occupies the lowest state outside the shell, and the other electron may occupy a variety of higher states. For this reason, the level is determined only by the quantum numbers n, ℓ of the electron in the higher state. In the case of Mg, the shell closes with $n = 2$, and therefore the quantum numbers for the lower electron are $n = 3$, $\ell = 0$, and those for the higher one can be $n = 3, 4, 5, \ldots, \ell = 0, 1, 2, 3, \ldots$ Accordingly as $\ell = 0, 1, 2, 3$, etc., the levels are named S, P, D, F, etc.

As I explained earlier, the doublet terms of alkalis exist because the spin s is parallel or antiparallel to the orbital angular momentum ℓ. Now we have the problem of triplet terms. Is it not possible that there exists a certain angular momentum not $|s| = 1/2$ but $|s| = 1$, and that it is azimuthally quantized with respect to the angular momentum ℓ?

In the core model of lecture 1, we considered this s as the angular momentum of the core. Its identity aside, $|s| = 1$ will be azimuthally quantized as sketched

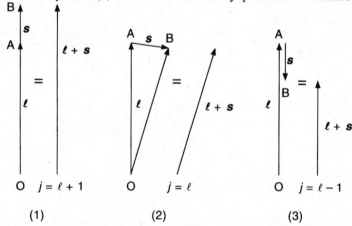

Figure 5.4 (1) s is parallel to ℓ, $|\ell + s| = \ell + 1$; (2) s is aligned such that $\overline{AO} = \overline{OB}$, $|\ell + s| = \ell$; (3) s is antiparallel to ℓ, $|\ell + s| = \ell - 1$.

in figure 5.4. When the angular momentum ℓ is not 0, in addition to parallel and antiparallel alignments there is the extra possibility of skew alignment, as you see in the figure. It is not clear just to say skew, but from the general requirement of quantum mechanics that the magnitude of angular momentum must always be integral or half-integral, its direction is determined. In our case $|s| = 1$, and we see from figure 5.4 that if we put

$$\ell + s = j, \tag{5-6}$$

then $|j|$, namely j, is

$$j = \begin{cases} \ell + 1 & \text{if } s \text{ and } \ell \text{ are parallel} \\ \ell & \text{if } s \text{ and } \ell \text{ are askew} \\ \ell - 1 & \text{if } s \text{ and } \ell \text{ are antiparallel}. \end{cases} \tag{5-7}$$

Thus there are three orientations of s with respect to ℓ, and from the analogy of alkali metals, the levels split into three. If we write the energy of the interaction between spin and orbital motion, we shall have, just like (5-4),

$$E_{s,\ell} = E_{s,\ell}(n, \ell; j). \tag{5-8}$$

The only difference from (5-4) is that the value of j is not given by (5-3) but by (5-7). (5-7) does not hold for the exceptional case of $\ell = 0$, when

$$j = 1. \tag{5-7'}$$

This corresponds to (5-3").

Then what is the angular momentum s for which $|s| = 1$? In the core model given in lecture 1, this was considered to be the angular momentum of the core, but with the advent of spin, it is superseded by the following theory. In the case of alkaline earths, there are two electrons outside the closed shell, and each electron has its own spin, which we write as s_1 and s_2. Of course, $s_1 = |s_1| = 1/2$, and $s_2 = |s_2| = 1/2$. We now assume that there is an interaction between the two spins which is stronger than $E_{s,\ell}$. Let us write this interaction between the spins as

$$E_{s,s} = E_{s,s}(n, \ell; s). \tag{5-9}$$

(We shall soon see that $E_{s,s}$ is a function of s as shown on the right-hand side.) Because this energy is larger than $E_{s,\ell}$, the two spins influence each other more strongly than they are influenced by ℓ. Therefore s_1 and s_2 will not be quantized with respect to ℓ but with respect to each other. Then s_1 and s_2 must be either mutually parallel or antiparallel. Therefore, the magnitude $s = |\mathbf{s}_1 + \mathbf{s}_2|$ of the added vector

$$\mathbf{s} = \mathbf{s}_1 + \mathbf{s}_2 \tag{5-10}$$

will be

$$s = \begin{cases} 1 & \text{if } \mathbf{s}_1 \text{ and } \mathbf{s}_2 \text{ are parallel} \\ 0 & \text{if } \mathbf{s}_1 \text{ and } \mathbf{s}_2 \text{ are antiparallel.} \end{cases} \tag{5-11}$$

(5-10) gives the s which we just considered. We can consider this to be s because this s can have a magnitude of 1.

Now if we look at (5-11), $s = 0$ is also possible. In this case, the added vector s is a zero vector, and the atom behaves as if s does not exist. So in this case $E_{s,\ell}$ also does not exist, and the energy levels of the atom will simply be those of the orbital motion. The levels then naturally are singlet. This argument also explains the singlet levels of Mg in figure 5.2.

Thus the energy levels of alkaline earths are

$$E_{\text{total}}(n, \ell; s; j) = E^{(1)}(n, \ell) + E_{s,s}(n, \ell; s) + E_{s,\ell}(n, \ell; s; j). \tag{5-12}$$

Here s in $E_{s,s}$ takes a value of 1 or 0. $E_{s,\ell} = 0$ if $s = 0$ or $\ell = 0$, and when $s = 1$, j in $E_{s,\ell}$ takes the three values $\ell + 1, \ell, \ell - 1$ as in (5-7). Therefore $E_{s,\ell}$ takes a different value for each value of j.

According to this idea, the singlet term of alkaline earths appears when the two spins are antiparallel, triplet states appear when they are parallel, and the spectrum can be neatly explained.

Now what is the order of magnitude of the spin-spin interaction $E_{s,s}$? We can answer this from the experimental data given in figure 5.2 by comparing the levels of the singlet terms and triplet terms. From the formula just obtained, (5-12), the difference between the energy of a singlet term with a certain n, ℓ and that of a triplet term with the same n, ℓ is

$$E_{s,s}(n, \ell; 0) - E_{s,s}(n, \ell; 1).$$

Therefore, if we obtain the interval by experiment, we can guess the magnitude of $E_{s,s}$. Then we find that the interaction between spins is quite large.

Now the problem is, What is the origin of this large $E_{s,s}$?

Certainly there is a magnetic interaction between $\mathbf{s}_1$ and $\mathbf{s}_2$. Since the electron has a magnetic moment of $-e\hbar/2mc$ along the direction of spin, there is an interaction energy between them of the order

$$E_{s,s} = \left(\frac{e\hbar}{2mc}\right)^2 \left\langle \frac{1}{|\mathbf{x}_1 - \mathbf{x}_2|^3} \right\rangle. \qquad (5\text{-}13)$$

However, such energy of magnetic origin causes a difference between $E_{s,s}$ $(n, \ell; 0)$ and $E_{s,s}(n, \ell; 1)$ of only the order of magnitude of splittings seen in alkali doublet levels. On the contrary, from figure 5.2 we see that the difference is as big as that of electric origin. This was an enigma for many years. However, if we assume this large $E_{s,s}$ without understanding its cause, we can explain a variety of complicated spectra. The answer to this question was found only after the discovery of the new quantum mechanics, especially the relation between the particle statistics and the symmetry of the wave function as I told you in the previous lecture. Let me explain this.

So how do we consider this in the new quantum mechanics? Before going into this problem, please note that the levels of Mg shown in figure 5.2 are all doubly degenerate according to the old idea. This is because if we exchange electrons 1 and 2 and consider electron 2 to be at the lowest level $n = 3$, $\ell = 0$ and electron 1 to be in a variety of n, ℓ states, we get exactly the same energy-level scheme.

Now let us go into the new quantum mechanics. We shall determine the levels in the following steps. In step (i) we solve the problem of two electrons without taking the spin degrees of freedom into account. This means we solve the Schrödinger equation with respect to the spatial motion of the electrons and determine its eigenvalues and eigenfunctions. In step (ii) we solve the Schrödinger equation for spin separately from the spatial motion and determine its eigenvalues and eigenfunctions. And finally in step (iii) we consider the interaction between spatial motion and spin. We shall do it this way. Now in this process something new appears which was totally unexpected in the old quantum mechanics. This is the issue which played an important role in the previous lecture, that is, whether the wave function is symmetric or antisymmetric with respect to the exchange of particles.

Let us first consider step (i). The problem here is to solve the Schrödinger equation

$$\left\{ \left[-\frac{\hbar^2}{2m}\Delta_1 + V_{Z=2}(|\mathbf{x}_1|) \right] + \left[-\frac{\hbar^2}{2m}\Delta_2 + V_{Z=2}(|\mathbf{x}_2|) \right] \right.$$
$$\left. + \frac{e^2}{|\mathbf{x}_1 - \mathbf{x}_2|} - E^{(1)} \right\} \psi(\mathbf{x}_1, \mathbf{x}_2) = 0. \qquad (5\text{-}14)$$

We now assume that we have solved the equation and obtained its eigenvalue $E^{(1)}$ and its eigensolution $\psi(\mathbf{x}_1, \mathbf{x}_2)$. If we interchange $\mathbf{x}_1$ and $\mathbf{x}_2$, then the expression inside braces does not change, but $\psi(\mathbf{x}_1, \mathbf{x}_2)$ becomes $\psi(\mathbf{x}_2, \mathbf{x}_1)$. Therefore, if $\psi(\mathbf{x}_1, \mathbf{x}_2)$ is an eigenfunction, then $\psi(\mathbf{x}_2, \mathbf{x}_1)$ must also be an eigenfunction. Furthermore, the latter also corresponds to the eigenvalue $E^{(1)}$. Therefore, if the eigenvalue $E^{(1)}$ is not degenerate (this will be shown later), $\psi(\mathbf{x}_2, \mathbf{x}_1)$ must be some constant times $\psi(\mathbf{x}_1, \mathbf{x}_2)$, namely

$$\psi(\mathbf{x}_2, \mathbf{x}_1) = \alpha \psi(\mathbf{x}_1, \mathbf{x}_2). \tag{5-15}$$

Now repeating the interchange of $\mathbf{x}_1$ and $\mathbf{x}_2$ produces a condition identical to noninterchange, and therefore α^2 must be 1, which means either

$$\psi(\mathbf{x}_2, \mathbf{x}_1) = \psi(\mathbf{x}_1, \mathbf{x}_2) \tag{5-16}_s$$

or

$$\psi(\mathbf{x}_2, \mathbf{x}_1) = -\psi(\mathbf{x}_1, \mathbf{x}_2). \tag{5-16}_a$$

That is to say, the eigenfunction $\psi(\mathbf{x}_1, \mathbf{x}_2)$ has to be either symmetric or antisymmetric with respect to the interchange of $\mathbf{x}_1$ and $\mathbf{x}_2$.

Now how can we assign quantum numbers to the eigenvalues and eigenfunctions of (5-14)?

We do it as follows. Namely, we examine how the level $E^{(1)}$ under consideration relates to $E^{(0)}$ when the interaction $e^2/|\mathbf{x}_1 - \mathbf{x}_2|$ is adiabatically set to zero, and we use the quantum numbers of $E^{(0)}$ as the quantum numbers of $E^{(1)}$. When we set $e = 0$ in this term of (5-14), we obtain

$$\left\{ \left[-\frac{\hbar^2}{2m}\Delta_1 + V_{Z=2}(|\mathbf{x}_1|) \right] + \left[-\frac{\hbar^2}{2m}\Delta_2 + V_{Z=2}(|\mathbf{x}_2|) \right] \right.$$
$$\left. - E^{(1)} \right\} \psi(\mathbf{x}_1, \mathbf{x}_2) = 0, \tag{5-14'}$$

which can be solved by the method of separation of variables. We can use the eigenvalues $E_{Z=2}(n_1, \ell_1)$, $E_{Z=2}(n_2, \ell_2)$ and the eigenfunctions $\psi_{n_1, \ell_1}(\mathbf{x}_1)$, $\psi_{n_2, \ell_2}(\mathbf{x}_2)$ of the one-electron Schrödinger equation

$$\left[-\frac{\hbar^2}{2m}\Delta_1 + V_{Z=2}(|\mathbf{x}_1|) - E_{Z=2}(n_1, \ell_1) \right] \psi_{n_1, \ell_1}(\mathbf{x}_1) = 0 \tag{5-17}_1$$

$$\left[-\frac{\hbar^2}{2m}\Delta_2 + V_{Z=2}(|\mathbf{x}_2|) - E_{Z=2}(n_2, \ell_2) \right] \psi_{n_2, \ell_2}(\mathbf{x}_2) = 0 \tag{5-17}_2$$

and obtain the eigenvalues and eigenfunctions of (5-14') as

$$E^{(0)}(n_1, \ell_1; n_2, \ell_2) = E_{Z=2}(n_1, \ell_1) + E_{Z=2}(n_2, \ell_2) \tag{5-18}$$

$$\psi^{(0)}_{n_1, \ell_1; n_2, \ell_2}(\mathbf{x}_1, \mathbf{x}_2) = \psi_{n_1, \ell_1}(\mathbf{x}_1)\psi_{n_2, \ell_2}(\mathbf{x}_2). \tag{5-19}$$

We can specify the quantum numbers $n_1, \ell_1; n_2, \ell_2$ for $E^{(1)}$ by observing how the eigenfunction of (5-14) relates to the eigenvalues and eigenfunctions (with their own n and ℓ) in (5-18) and (5-19) when (5-14) is adiabatically brought to (5-14').

Now one problem occurs here. The problem is that although the eigenfunction of (5-14) has to be either symmetric or antisymmetric with respect to the interchange of $\mathbf{x}_1$ and $\mathbf{x}_2$, (5-19) is neither symmetric nor antisymmetric unless $n_1 = n_2$ and $\ell_1 = \ell_2$. Therefore, the eigenfunction ψ of (5-14) cannot be correlated to (5-19) unless $n_1 = n_2$ and $\ell_1 = \ell_2$.

The fact that eigensolutions exist which are neither symmetric nor antisymmetric indicates that the assumption which we used in deriving (5-16)$_\text{s}$ and (5-16)$_\text{a}$, namely that the eigenvalues are not degenerate, is not valid. Indeed, if we interchange electrons 1 and 2, we obtain instead of (5-18)

$$E^{(0)}(n_2, \ell_2; n_1, \ell_1) = E_{Z=2}(n_2, \ell_2) + E_{Z=2}(n_1, \ell_1), \tag{5-18'}$$

but this has a value identical to that of (5-18). On the other hand, for the eigenfunction we obtain

$$\psi^{(0)}_{n_2, \ell_2; n_1, \ell_1}(\mathbf{x}_1, \mathbf{x}_2) = \psi_{n_2, \ell_2}(\mathbf{x}_1)\psi_{n_1, \ell_1}(\mathbf{x}_2), \tag{5-19'}$$

and this does not agree with (5-19).

However, if we make the linear combinations

$$\psi^{(0)\text{sym}}_{n_1, \ell_1; n_2, \ell_2} = \psi_{n_1, \ell_1}(\mathbf{x}_1)\psi_{n_2, \ell_2}(\mathbf{x}_2) + \psi_{n_2, \ell_2}(\mathbf{x}_1)\psi_{n_1, \ell_1}(\mathbf{x}_2) \tag{5-20$_\text{s}$}$$

and

$$\psi^{(0)\text{ant}}_{n_1, \ell_1; n_2, \ell_2} = \psi_{n_1, \ell_1}(\mathbf{x}_1)\psi_{n_2, \ell_2}(\mathbf{x}_2) - \psi_{n_2, \ell_2}(\mathbf{x}_1)\psi_{n_1, \ell_1}(\mathbf{x}_2), \tag{5-20$_\text{a}$}$$

then $\psi^{(0)\,\text{sym}}$ is symmetric and $\psi^{(0)\,\text{ant}}$ is antisymmetric, so if $\psi(\mathbf{x}_1, \mathbf{x}_2)$ is symmetric, it correlates to (5-20)$_\text{s}$, and if $\psi(\mathbf{x}_1, \mathbf{x}_2)$ is antisymmetric, it correlates to (5-20)$_\text{a}$, although the eigenfunction of (5-14) is neither (5-19) nor (5-19'). Even in the old quantum mechanics, in general we cannot literally say that electron 1 is in the state n_1, ℓ_1 and electron 2 is in n_2, ℓ_2 when there is an interaction between them, but if the interaction goes to 0, we can regard them in that way. Now in the new quantum mechanics, we cannot say this even if the interaction is very small, even 0, and the electron is always in a superposition of the state with $n_1, \ell_1; n_2, \ell_2$ and the state with $n_2, \ell_2; n_1, \ell_1$. This concept of the superposition of states exists only in the new quantum mechanics, and there was absolutely no such idea in classical mechanics.

So now we see that if we set the interaction to 0, then the state with eigenvalue

$$E^{(0)}(n_1, \ell_1; n_2, \ell_2) = E_{Z=2}(n_1, \ell_1) + E_{Z=2}(n_2, \ell_2) \qquad (5\text{-}20)$$

is doubly degenerate, and one eigenvalue has eigenfunction $(5\text{-}20)_s$ and the other $(5\text{-}20)_a$. Now let us consider how this $E^{(0)}$ varies when we introduce the interaction into the system adiabatically. We can immediately understand without calculation that $\psi^{(0)\text{sym}}$ of $(5\text{-}20)_s$ is nonzero when $\mathbf{x}_1 = \mathbf{x}_2$, whereas $\psi^{(0)\text{ant}}$ of $(5\text{-}20)_a$ is always zero when $\mathbf{x}_1 = \mathbf{x}_2$. From this we can conclude, roughly speaking, that the probability of two electrons being close is smaller in the state ψ^{ant} than in the state ψ^{sym}. Then since there will be a repulsive force between the two electrons, the energy will be lower if they do not approach closely. ψ^{ant} will have lower energy than ψ^{sym}. In this case, the difference of the two levels has its origin in the electric Coulomb interaction and therefore should be rather large. Therefore, if we change e in the numerator of $e^2/|\mathbf{x}_1 - \mathbf{x}_2|$ gradually from 0 to 4.8×10^{-10} esu, then the repulsive potential gradually comes in, and all levels gradually increase from the values given by $(5\text{-}20)$, but at the same time they split, and the splitting becomes gradually larger with the ψ^{ant} level below the ψ^{sym} level. We can write the variation of the energy levels due to the electron repulsion as

$$\left\langle \frac{e^2}{|\mathbf{x}_1 - \mathbf{x}_2|} \right\rangle^{\text{sym}}_{n_1, \ell_1; n_2, \ell_2} \qquad (5\text{-}21)_s$$

for the symmetric state and

$$\left\langle \frac{e^2}{|\mathbf{x}_1 - \mathbf{x}_2|} \right\rangle^{\text{ant}}_{n_1, \ell_1; n_2, \ell_2} \qquad (5\text{-}21)_a$$

for the antisymmetric state. Then in general, we can show that the former is larger than the latter and that the eigenvalues of $(5\text{-}14)$ are nondegenerate.

From the foregoing discussion we now know how to assign quantum numbers $n_1, \ell_1; n_2, \ell_2$ to eigenvalues and eigenfunctions of $(5\text{-}14)$. We compared which $n_1, \ell_1; n_2, \ell_2$ correlate when e in the numerator of $e^2/|\mathbf{x}_1 - \mathbf{x}_2|$ is set equal to zero and deduced the symmetry property of ψ. Here we have to recognize that the order of n_1, ℓ_1 and n_2, ℓ_2 is meaningless because both in $(5\text{-}20)_s$ and in $(5\text{-}20)_a$ the order of these is meaningless. Therefore, the eigenvalues of (5-14) can be written as

$$E^{(1)\text{sym}}_{n_1,\ell_1;n_2,\ell_2}, \qquad E^{(1)\text{ant}}_{n_1,\ell_1;n_2,\ell_2} \qquad (5\text{-}22)$$

and their eigenfunctions as

$$\psi^{(1)\text{sym}}_{n_1,\ell_1;n_2,\ell_2}(\mathbf{x}_1, \mathbf{x}_2), \qquad \psi^{(1)\text{ant}}_{n_1,\ell_1;n_2,\ell_2}(\mathbf{x}_1, \mathbf{x}_2). \qquad (5\text{-}23)$$

Now we start comparing with experimental observations; we shall take as an example the ^1P and ^3P levels of Mg in figure 5.2. For those, we can set $n_1 = 3$, $\ell_1 = 0$ and $n_2 = n$, $\ell_2 = 1$. This means one of the electrons is in the lowest state outside the closed shell, and the other electron will be in the P state with various n values.

I summarize this situation in figure 5.5. As I just told you, I consider only $n_1 = 3, \ell_1 = 0, n_2 = n, \ell_2 = 1$. However, as I told you earlier, the state in question must be a superposition of $n_1 = 3, \ell_1 = 0, n_2 = n, \ell_2 = 1$ and $n_1 = n, \ell_1 = 1, n_2 = 3, \ell_2 = 0$. We first write in the leftmost column $E^{(0)}$ of (5-20) with $n_1 = 3, \ell_1 = 0, n_2 = n, \ell_2 = 1$. Since the level can be specified only by n and $\ell_2 = 1$, I omitted $n_1 = 3, \ell_1 = 0$. Namely, instead of $E^{(0)}(3, 0, n, 1)$ I simply write $E^{(0)}(n, 1)$. Next we take into account $e^2|\mathbf{x}_1 - \mathbf{x}_2|$. Then as we saw near the end of the previous section, the levels split into two, and as the value of e is increased, the value of the splitting gradually increases with the antisymmetric level always below the symmetric level. In order to show whether levels are symmetric or antisymmetric with respect to the interchange of $\mathbf{x}_1$ and $\mathbf{x}_2$, I labeled them x-sym and x-ant. Just as in the old theory, the $E^{(0)}$ levels are doubly degenerate, so I drew them as bold lines. However, the levels $E^{(1)}(n, 1)$ split into two, x-sym and x-ant, and are no longer degenerate. Therefore, I do not use bold lines for the $E^{(1)}$ levels.

Now we go on to step (ii), which is the inclusion of the spin degree of freedom. As I told you in the third lecture, related to Pauli's spin theory, it is convenient to use for the spin coordinate in the wave function the z-component s_z. If we write the spin coordinate of the first electron as s_{1z} and that of the second as s_{2z}, then we can write the wave function with the spin degrees of freedom as

$$\psi(s_{1z}, s_{2z}). \qquad (5\text{-}24)$$

Although the value of total spin $\mathbf{s} = \mathbf{s}_1 + \mathbf{s}_2$ is 0 or 1 in the new quantum mechanics as in the old quantum mechanics, the new concept that did not exist in the old quantum mechanics is the idea of the symmetry of $\psi(s_{1z}, s_{2z})$ upon

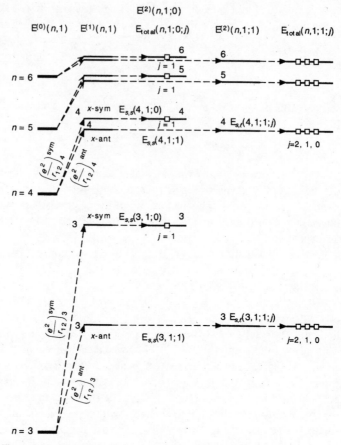

Figure 5.5 Interpretation of spectral terms by new quantum mechanics.

interchange of the coordinates s_{1z} and s_{2z}. Please remember the theorem which I proved in the third lecture. The theorem was that if the magnitude of the total spin $\mathbf{s}$ is s and the corresponding wave function is $\psi_s(s_{1z}, s_{2z})$, then

$$\psi_s(s_{1z}, s_{2z}) \text{ is } \begin{cases} \text{symmetric if } s = 1 \\ \text{antisymmetric if } s = 0. \end{cases} \qquad (5\text{-}24')$$

(The converse of this theorem is also true.) Unlike in the old theory, we do not have to consider energies other than the interaction energy between the magnetic moment of the electrons to consider the energies of $\psi^{(1)}$ and $\psi^{(0)}$. If this is the case, the contribution to the energy level from (5-13) is only on the order of

$$E_{s,s} = \left(\frac{e\hbar}{2mc}\right)^2 \left\langle \frac{1}{|\mathbf{x}_1 - \mathbf{x}_2|^3} \right\rangle_{n,\ell;s}. \tag{5-25}$$

This is very small both for $s = 0$ and $s = 1$ and is almost negligible. Therefore, $E^{(2)}(n, 1; s)$ for all practical purposes can be approximated as $E^{(1)}(n, 1)$ regardless of the value of s.

The reason we cited theorem (5-24') is that there is a specific relation between the particle's statistics and the symmetry of the wave function, which we must take into account here. Namely, the total wave function, which contains both spatial and spin coordinates of the electron, is given by the product of a function with spatial degrees of freedom like (5-23) and a function of spin degrees of freedom like (5-24'); since the electron is a fermion, this total function must be antisymmetric with respect to the interchange of $\mathbf{x}_1, s_{1z}$ and $\mathbf{x}_2, s_{2z}$. From this requirement the following unfolds. By using the same argument with which I derived (4-9), the total wave functions

$$\psi^{(1)\text{sym}}_{n_1,\ell_1;n_2,\ell_2}(\mathbf{x}_1, \mathbf{x}_2) \cdot \psi_0(s_{1z}, s_{2z}) \tag{5-26}$$

and

$$\psi^{(1)\text{ant}}_{n_1,\ell_1;n_2,\ell_2}(\mathbf{x}_1, \mathbf{x}_2) \cdot \psi_1(s_{1z}, s_{2z}) \tag{5-27}$$

are both antisymmetric with respect to the interchange of $\mathbf{x}_1, s_{1z}$ and $\mathbf{x}_2, s_{2z}$, and these states can actually occur, but

$$\psi^{(1)\text{ant}}_{n_1,\ell_1;n_2,\ell_2}(\mathbf{x}_1, \mathbf{x}_2) \cdot \psi_0(s_{1z}, s_{2z}) \tag{5-26'}$$

$$\psi^{(1)\text{sym}}_{n_1,\ell_1;n_2,\ell_2}(\mathbf{x}_1, \mathbf{x}_2) \cdot \psi_1(s_{1z}, s_{2z}) \tag{5-27'}$$

are both symmetric with respect to the interchange of $\mathbf{x}_1, s_{1z}$ and $\mathbf{x}_2, s_{2z}$, and thus these states cannot exist in reality.

Therefore, when spin degrees of freedom are taken into account, from the levels in the second column of figure 5.5, only the $s = 0$ level results from the x-sym level, and only $s = 1$ results from the x-ant level. Therefore if we write in the third column $E^{(2)}(n, 1; 0)$ and in the fourth column $E^{(2)}(n, 1; 1)$, then since $E_{s,s}$ is very small, the x-sym of $E^{(1)}(n, 1)$ moves to the third column, and x-ant slides to the fourth column.

Finally, we go on to step (iii). Here we incorporate the spin-orbit interaction energy $E_{s,\ell}$ to obtain E_{total}. The result is just like that as before: the $E^{(2)}$ level of $s = 0$ does not split, but $E^{(2)}$ of $s = 1$ splits into three. We again express the states by the notation ☐ and ☐☐☐. Then we get the columns numbered 3 and 5 in figure 5.5.

For comparison with figure 5.5, I give the procedure of the old quantum mechanics in figure 5.6. I do not want to go into detail, but in this procedure the splitting of singlet and triplet appears when $E_{s,s}$ is introduced. Also, in the old theory all the levels are doubly degenerate, and therefore all of them in the figure are shown by bold lines. As these two figures indicate, the thoughts underlying the two level schemes are quite different. In the old theory the splitting between singlet and triplet occurs first when $E_{s,s}$ is taken into account, but in the new theory it already appears when we move from $E^{(0)}$ to $E^{(1)}$ through the contribution of $\langle e^2/|\mathbf{x}_1 - \mathbf{x}_2|\rangle$. In the new theory one of the split levels is singlet and the other is triplet *because* the electron is a fermion. In the old theory

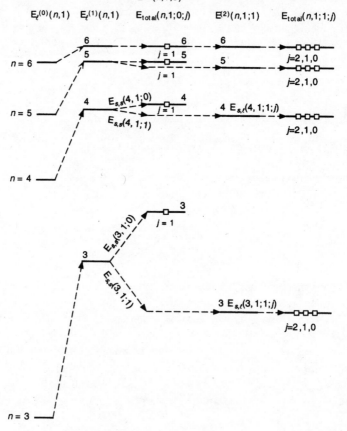

Figure 5.6 Interpretation of spectral terms by old quantum mechanics.

we had to consider a large spin-spin interaction $E_{s,s}$, while in the new theory it is sufficient to consider a small $E_{s,s}$ of magnetic origin. Also as I mentioned earlier, in the old theory all the levels are doubly degenerate, but not any more in the new theory. This is because the states given by (5-26') and (5-27') must be dropped from consideration.

Thus we have answered the puzzle of atomic spectra as to why electron spins interact with each other on such a large scale, unthinkable for a magnetic interaction. When Pauli negated the core model and proposed the two-valuedness of the electron, he left for the future why the alkaline earth spectrum splits into singlet and triplet and what causes the energy difference; this question has now been answered as due to the symmetry property of the wave function, which had not been considered at all at that time.

Here we considered ^1P terms and ^3P terms, but you can do the same for ^1S and ^3S terms. You will obtain the conclusion that in ^3S terms the $n = 3$ term must be dropped. This last conclusion was verified by Pauli using the exclusion principle, but here we go one step deeper and discover through this exercise that the reason for the Pauli exclusion principle is the antisymmetric property of the wave function.

The effect of the apparently large interaction between electron spins is not limited to spectral term values. In order to explain the ferromagnetism of Fe, as you may know, P. Weiss proposed long ago that there is a large interaction between molecular magnets based on the then-accepted concept of the molecular magnet. By using this idea Weiss could explain a wide variety of experimental results related to ferromagnetism. However, the origin of such a strong interaction between molecular magnets was entirely unknown.

Then there appeared the new interpretation of the spectral terms of alkaline earths. This new interpretation was given by Heisenberg in 1926; he not only discovered that the symmetry property of the wave function has a close connection to a particle's statistics in a many-electron system but also found that it plays an important role in a variety of problems and for the first time gave a clear explanation of the spectral terms of two-electron systems. Furthermore, immediately after this work he applied the same idea to the problem of ferromagnetism.

I have been talking for some time now so I would not proceed any further on the problem of ferromagnetism, but ferromagnetism occurs because electron spins outside all the Fe atomic cores point in the same direction over an entire macroscopic crystal of Fe. The apparently very strong spin-spin interaction just discussed plays the role of orienting spins in the same direction. Therefore, the subtle property of electron spin appears directly in the everyday macroscopic phenomenon of a magnet attracting iron, and this is related to wholly transcendent facts such as the symmetry property of wave functions and the electron being a fermion. This is one good example of transcendent theory appearing directly in ordinary, everyday phenomena. If the reason for ferromagnetism is the alignment of spins on a macroscopic scale, then not only the total magnetic moment but also the total spin angular momentum should have a macroscopic value. If so, in order to satisfy the conservation of angular momentum when a body magnetizes, the body should acquire an angular momentum opposite to the total spin angular momentum and should start to rotate, and this rotation should be observable. Actually, Einstein and W. J. de Haas had already demonstrated this experimentally in 1915 and from the rotation had measured $g\left(\frac{\text{component of magnetic moment along the magnetic field}}{\text{component of angular momentum along the magnetic field}}\right)$ of Fe. In this experiment they obtained a value for g which is close to 1, but later many people did the experiment with higher accuracy and obtained $g = g_0 = 2$ not only for Fe but also for Ni and Co. These experiments show clearly that the origin of the magnetization of a ferromagnetic material is the electron spin.

Let me finally add one thing. After Heisenberg's paper on alkaline earths, Dirac also published a paper on the apparent spin-spin interaction. In this paper he derives the spin-spin interaction ingeniously by considering the interchange of particles (in general, permutation) as a physical quantity, an idea that would not occur to mortals. I shall not go further into this theory because it is described in detail in Dirac's textbook.

Pauli-Weisskopf and the Yukawa Particle

There Is No Reason for Nature to Reject Particles with Spin Zero

With this lecture I return to the main stream which was previously discussed. In the third lecture I described "why nature is not satisfied by a simple point charge." Dirac answered this question. Maybe I am too gossipy, but I suspect that when Pauli saw Dirac's work, he thought he had been scooped by him. Pauli himself was trying to incorporate spin into quantum mechanics and introduced the Pauli matrices, but he could not arrive at the complete theory. He also clearly recognized the necessity of making the theory relativistic and yet could not achieve it. He was known to be a perfectionist who severely criticized other people's work, but he had no alternative other than to publish his unsatisfactory ad hoc theory. Now on the other hand, Dirac established his theory with totally unexpected acrobatics and solved all the problems on spin as well as making the theory relativistic. I do not believe Pauli could have been so ascetic as to have accepted this graciously.

However, in 1934 Pauli ended up retaliating against Dirac. Pauli showed that Dirac's argument that nature is satisfied only by the Dirac equation and not by the Klein-Gordon equation is incorrect. According to him, the Klein-Gordon equation is not in contradiction with the framework of quantum mechanics, and there is no reason nature abhors particles with spin 0.

I would like to talk today on how this came about, but we need quite a bit of preparation. The first item is the quantization of the wave field, and the other is the problem of Dirac's negative energy.

Probably you have had the same experience, but the "wave," when we study wave mechanics or the wave function ψ, is discussed sometimes as if it were a real wave in the three-dimensional space in which we live and sometimes as if it were an abstract wave in the configuration space. Have you ever worried as to which is the case? Such confusion with the concept occurred quite a bit

Figure 6.1 Wolfgang Pauli, 1940
passport photo. [Courtesy of AIP
Emilio Segrè Visual Archives]

at the beginning of the development of quantum mechanics. Indeed, if we look at Schrödinger's series of papers "Quantization as an Eigenvalue Problem," we find that Schrödinger himself was going back and forth between these two ideas. However, by and large he tended to think that his ψ was a wave in three-dimensional space. For example, he considered $e\psi^*(\mathbf{x})\psi(\mathbf{x})$ as the charge density which actually exists in space and tried to treat the bulk of the density as an electron. This idea, however, did not work because $\psi^*\psi$ will spread with time and the density becomes diffuse.

On the other hand, when quantum mechanics was completed in the form of transformation theory, Schrödinger's ψ was established as a representation of the state vector of the mechanical system. It was also called the probability amplitude, which is a function of the generalized coordinates describing the mechanical system, and therefore the wave expressed by ψ exists in the abstract coordinate space and not in our three-dimensional space. Therefore, for two particles, ψ is a function $\psi(\mathbf{x}_1, \mathbf{x}_2)$ of the coordinate $\mathbf{x}_1$ of the first particle and the coordinate $\mathbf{x}_2$ of the second particle, and this is a wave in a six-dimensional space. Furthermore, more generally, the variables which constitute the argument of ψ are not limited to the spatial coordinates $\mathbf{x}_1, \mathbf{x}_2, \ldots$ but could be the generalized coordinates of Lagrange or even the more abstract canonical coordinates of Hamilton's mechanics.

Therefore, it looks as if Schrödinger's idea of a wave of real matter seems to have been completely superseded. The idea that the radiation wave does indeed exist in our three-dimensional space but the matter wave does not was about to become orthodox.

However, this situation suddenly changed when the idea of quantization of the wave field was introduced. Namely, it was established that the concept of a matter wave actually existing in space is just as valid as that of a light wave existing in space.

Quantum mechanics was initially developed to study material particles such as electrons and nuclei. And the phenomena of atoms emitting, absorbing, and scattering radiation were treated through the analogy with classical mechanics with the assumption that the matrix elements of the electronic coordinates correspond to the Fourier components of the classical electron motion; there was no other recourse than to treat the whole problem by analogy to classical theory. However, when quantum mechanics was completed in the form of transformation theory, it was ready to be applied to the radiation field. There was an attempt to consider a mechanical system that includes atoms as well as a radiation field and to apply quantum mechanics also to the field, treating the interaction of radiation and matter consistently. It was Dirac who initiated this attempt in 1927.

The idea of treating the radiation field quantum mechanically was by no means new. Already at the beginning of quantum theory was Debye's idea that if you decompose the radiation field into plane waves and make its amplitude discrete such that it satisfies Planck's condition

$$E_\nu = N_\nu h\nu, \qquad N_\nu = 0, 1, 2, \ldots,$$

then you can derive Planck's formula. Therefore, instead of simply using Planck's condition ad hoc, we can reconsider the amplitude of the wave as a matrix (or Dirac's q-number) and apply the new quantum mechanics. Then the condition $E_\nu = N_\nu h\nu$ will be naturally derived, and we can interpret N_ν as the number of photons with energy $h\nu$. If we do this, the particle nature of radiation will automatically emerge. Indeed soon after Heisenberg devised matrix mechanics, Born, Jordan, and Heisenberg, who completed matrix mechanics, themselves proposed to consider the electric and magnetic fields $\mathbf{E}$ and $\mathbf{H}$ also as matrices (although at that time the idea of probability amplitude did not yet exist, and they could not apply their idea to the emission, absorption, and scattering of radiation).

Needless to say, Dirac naturally started from the idea of Debye and Born. This initial step could have been taken by any of us mortals (regardless of whether we can complete it or not), but Dirac introduced one further idea which is characteristic of him. It is the weird idea which later came to be known as *second quantization*. I am afraid that I must digress from spin, but this weird idea is very typical of Dirac, and with this idea the *matter wave* is incorporated into the three-dimensional space, so let me spend some time explaining it.

Dirac first considered the Schrödinger equation for one particle. As you know, it is

$$\left[H(\mathbf{x}, \mathbf{p}) + \frac{\hbar}{i} \frac{\partial}{\partial t} \right] \psi(\mathbf{x}, t) = 0 \tag{6-1}$$

$$H(\mathbf{x}, \mathbf{p}) = \frac{1}{2m} \mathbf{p}^2 + V(\mathbf{x}), \qquad \mathbf{p} = \frac{\hbar}{i} \nabla.$$

According to his transformation theory, $\psi(\mathbf{x}, t)$ represents the state vector of the system at time t. Now we consider some mechanical quantity (in Dirac's nomenclature, an observable) $G(\mathbf{x}, \mathbf{p})$, and let us call its eigenvalue g_n and eigenfunction $\phi_n(\mathbf{x})$. Here, $n = 1, 2, 3, \ldots$ Since the $\phi_n(\mathbf{x})$ constitute a complete, orthogonal system, we can expand $\psi(\mathbf{x}, t)$ as

$$\psi(\mathbf{x}, t) = \sum_n a_n(t) \phi_n(\mathbf{x}). \tag{6-2}$$

Then if we measure the observable $G(\mathbf{x}, \mathbf{p})$ at time t, the probability of obtaining the value g_n is

$$P_n(t) = |a_n(t)|^2. \tag{6-3}$$

This was a conclusion from the transformation theory. Here ψ and ϕ_n are all normalized to 1. $H(\mathbf{x}, \mathbf{p})$ represents the energy of the system, and in terms of the matrix elements of H,

$$H_{n,n'} = \int \phi_n^*(\mathbf{x}) H(\mathbf{x}, \mathbf{p}) \phi_{n'}(\mathbf{x}) dv, \tag{6-4}$$

the expectation value of H is given by

$$\langle H \rangle \equiv \int \psi^*(\mathbf{x}, t) H(\mathbf{x}, \mathbf{p}) \psi(\mathbf{x}, t) dv \tag{6-5}$$

$$= \sum_{n,n'} a_n^* H_{n,n'} a_{n'}.$$

Here it is easy to see that $\langle H \rangle$ is independent of time.

Using the matrix element (6-4), we can derive the time dependence of $a_n(t)$ as

$$\frac{da_n(t)}{dt} = \frac{1}{i\hbar} \sum_{n'} H_{n,n'} a_{n'}(t) \tag{6-6}$$

from (6-1). Its complex conjugate becomes

$$\frac{da_n^*(t)}{dt} = -\frac{1}{i\hbar} \sum_{n'} a_{n'}^*(t) H_{n',n}. \qquad (6\text{-}6)^*$$

If we examine (6-6) and (6-6)*, we see that we can rewrite them using $\langle H \rangle$ of (6-5), namely

$$\frac{da_n}{dt} = \frac{1}{i\hbar} \frac{\partial \langle H \rangle}{\partial a_n^*}, \qquad \frac{da_n^*}{dt} = -\frac{1}{i\hbar} \frac{\partial \langle H \rangle}{\partial a_n}. \qquad (6\text{-}7)$$

From these, if we treat a_n like the coordinate variable and

$$\pi_n = i\hbar a_n^* \qquad (6\text{-}8)$$

like its conjugate momentum, and consider

$$\langle H \rangle = \frac{1}{i\hbar} \sum_{n,n'} \pi_n H_{n,n'} a_n' \qquad (6\text{-}9)$$

as the Hamiltonian, then we can show that a_n and π_n satisfy the canonical equations of motion

$$\frac{da_n}{dt} = \frac{\partial \langle H \rangle}{\partial \pi_n}, \qquad \frac{d\pi_n}{dt} = -\frac{\partial \langle H \rangle}{\partial a_n}. \qquad (6\text{-}10)$$

So far we have considered one mechanical system consisting of one particle, but let us follow Dirac here and consider an ensemble composed of N mechanical systems each containing one particle. If we do so, the expectation value of the number of systems in the ensemble which have the value g_n for the observable G at one time is given as $N P_n$, namely

$$N_n \equiv N P_n = N |a_n|^2. \qquad (6\text{-}3')$$

Therefore, if we set

$$A_n = N^{1/2} a_n \quad \text{and} \quad A_n^* = N^{1/2} a_n^*, \qquad (6\text{-}11)$$

then we get

$$N_n = A_n^* A_n, \qquad (6\text{-}3'')$$

and if we set, corresponding to (6-8),

$$\Pi_n = i\hbar A_n^*, \tag{6-8'}$$

and, corresponding to (6-9),

$$\overline{H} = \frac{1}{i\hbar} \sum_{n,n'} \Pi_n H_{n,n'} A_{n'}, \tag{6-9'}$$

then we obtain, corresponding to (6-10),

$$\frac{dA_n}{dt} = \frac{\partial \overline{H}}{\partial \Pi_n}, \qquad \frac{d\Pi_n}{dt} = -\frac{\partial \overline{H}}{\partial A_n}. \tag{6-10'}$$

If we examine (6-10'), here also we can regard A_n and Π_n as canonical variables, which satisfy canonical equations with $\overline{H}$ as the Hamiltonian. We have the relation

$$\sum_n |a_n|^2 = \int \psi^*(\mathbf{x}, t)\psi(\mathbf{x}, t)dv = 1 \tag{6-12}$$

for a_n, while for A_n,

$$\sum_n |A_n|^2 = N. \tag{6-12'}$$

Dirac says that if you worry about the fact that A_n and Π_n are complex numbers, then use N_n and Θ_n defined by

$$A_n = N_n^{1/2} e^{i\Theta_n/\hbar} \qquad A_n^* = N_n^{1/2} e^{-i\Theta_n/\hbar}. \tag{6-13}$$

Dirac proved that these N_n and Θ_n are canonical variables conjugate to each other. When we use these variables, the Hamiltonian is

$$\overline{H} = \sum_{n,n'} N_n^{1/2} e^{-i\Theta_n/\hbar} H_{n,n'} N_{n'}^{1/2} e^{i\Theta_{n'}/\hbar}. \tag{6-9''}$$

Here Dirac performed his characteristic acrobatics. Namely, he redefined these A_n and Π_n as quantum mechanical q-numbers rather than as ordinary numbers. That is, he introduced the canonical commutation relation between A_n and Π_n to quantize the problem,

$$A_n \Pi_{n'} - \Pi_{n'} A_n = i\hbar \delta_{nn'}, \tag{6-14}$$

$$A_n A_{n'} - A_{n'} A_n = \Pi_n \Pi_{n'} - \Pi_{n'} \Pi_n = 0.$$

Why do I call this acrobatics? After all, the Schrödinger equation (6-1) is already the result of quantization. Therefore, all of equations (6-10) and (6-10') derived from that are already quantized. Why must you quantize it once more as the name second quantization suggests? We mortals stand bewildered here. However, there is no use being bewildered, so let us try to discover why we feel bewildered. We shall find the following.

The quantities which are expressed by q-numbers in quantum mechanics, be they the coordinate q, the momentum p, the energy H, or the quantity G that we just discussed, are all observables. The concept of observable was introduced by Dirac, and these are the quantities which are directly measurable by some experiments. However, quantities like Π_n, defined in (6-3), are nothing like that. In order to determine them, we repeat the measurement of observable G many times, accumulate a body of data, and then examine which fraction of measurements give g_n. Therefore Π_n is not a quantity which is directly measurable by experiment, and therefore it is not an observable. In that sense, none of a_n, a_n^*, π_n are observables.

We can say the same thing about N_n, A_n, A_n^*, and Π_n. First of all, when I said "an ensemble composed of N mechanical systems each with one particle," I meant by *ensemble* an imagined ensemble, and there is no need to have a real mechanical system composed of N particles in front of you. We can consider one mechanical system and repeat the measurement of G many times under the same conditions and then obtain the number of times N_n in which the value of G was measured to be g_n. Or we can make the mechanical system number one on the first day and measure G and destroy the system after the measurement, make a second mechanical system on the second day and measure the value of G and destroy that system, make a third mechanical system on the third day, and so on. Therefore, the ensemble we are considering here is a so-called virtual ensemble in statistics. When I say number of measurements N_n, it is not a quantity which itself is obtained by measuring an observable, but it is a number related to the accumulation of data on many repeated measurements of observables.

The reason why we were bewildered by Dirac's second quantization, which regards N_n, A_n, and Π_n as q-numbers, lies right here. Some of you might accept the second quantization without any difficulty. Such a person has to be either as great as Dirac or a happy-go-lucky type who, although he does not consider anything deeply, feels as if he understands everything.

Then on what basis did Dirac dare to quantize N_n, A_n, and Π_n in spite of these problems? I think the answer is as follows.

We know the fact that the statistical conclusion for virtual ensembles often agrees with that for real ensembles. Therefore, we may well expect that in the present problem there may also be such agreement. If so, we can apply our conclusion for a "virtual ensemble composed of N mechanical systems each with one particle" to the "mechanical system of N particles which are not mutually interacting." Now in this real ensemble we can consider N_n as an observable because the determination of N_n for a mechanical system composed of N particles amounts to the measurement of G for particle 1, 2, 3, ..., N, i.e., $G(\mathbf{x}_1, \mathbf{p}_1)$, $G(\mathbf{x}_2, \mathbf{p}_2)$, $G(\mathbf{x}_3, \mathbf{p}_3)$, ..., $G(\mathbf{x}_N, \mathbf{p}_N)$. Here the G of the N particles all commute, and therefore according to the transformation theory, we can measure all N of the G's simultaneously and regard it as one set of measurements. Furthermore, observable N_n is not an analytic function but is a function of N observables G (since the value of N_n is determined when the NG values which mutually commute are measured, N_n is a function of all the NG's), and therefore N_n can be determined by one set of measurements. For this reason we can consider N_n as a q-number in our real ensemble and likewise A_n and Π_n.

For these reasons, what Dirac did was a heuristic study to find the answer to the question whether it is possible to describe a many-particle system using the q-numbers N_n and their conjugates Θ_n or the q-numbers A_n and Π_n, and for that purpose to use equation (6-10') as the fundamental equation starting from the Hamiltonian (6-9'). Namely, he is using the second quantization as a heuristic means to see if such an approach is possible and only in this context.

Since the second quantization is a heuristic logic, we must examine whether the theory using the Hamiltonian (6-9') or (6-9") indeed agrees with the conclusion obtained from the treatment of many-particle systems in the ordinary way. [In the process of quantizing (6-9"), the order of $N_n^{1/2}$ and $e^{\pm i\Theta_n/\hbar}$ must be considered, but the right order will be given later in (6-17').] Here by ordinary way I mean the method of considering ψ in coordinate space which for the N particles gives the Schrödinger equation

$$\left[H(\mathbf{x}_1, \mathbf{p}_1) + H(\mathbf{x}_2, \mathbf{p}_2) + \ldots + H(\mathbf{x}_N, \mathbf{p}_N) + \frac{\hbar}{i} \frac{\partial}{\partial t} \right] \cdot$$
$$\psi(\mathbf{x}_1, \mathbf{x}_2, \ldots, \mathbf{x}_N) = 0. \qquad (6\text{-}15)$$

Dirac actually gives in his paper a proof that this agreement does result if he takes only the solution $\psi(\mathbf{x}_1, \mathbf{x}_2, \ldots, \mathbf{x}_n)$ which is symmetric with respect to the interchange of particles. By this procedure it has been demonstrated that the solution of the problem by using Hamiltonian (6-9') and commutation relation (6-14) is equivalent to solving (6-15) for an N-boson system with the Hamiltonian

$$H = \sum_{\nu=1}^{N} H(\mathbf{x}_\nu, \mathbf{p}_\nu). \qquad (6\text{-}16)$$

Here the probability amplitude that appears when we use Hamiltonian (6-9') or (6-9") is a function of the type of coordinate used, namely A_n or N_n

$$\psi(A_1, A_2, \ldots, A_n, \ldots) \quad \text{or} \quad \psi(N_1, N_2, \ldots, N_n, \ldots). \quad (6\text{-}17)$$

In particular, the latter is more commonly used, and the Schrödinger equation for that case is

$$\left[\sum_{n,n'} N_n^{1/2} e^{-i\Theta_n/\hbar} H_{n,n'} e^{i\Theta_{n'}/\hbar} N_{n'}^{1/2} + \frac{\hbar}{i} \frac{\partial}{\partial t} \right] \cdot$$

$$\psi(N_1, N_2, \ldots, N_n, \ldots) = 0. \quad (6\text{-}17')$$

Dirac showed that the operator $e^{\pm i\Theta/\hbar}$ in this equation has the property

$$e^{\pm i\Theta/\hbar} \psi(N) = \psi(N \pm 1).$$

Actually this proof was not rigorous, but it is very Dirac-like and very interesting as heuristic theory, so let me introduce it here.

Proof: Since Θ is the momentum conjugate to N, we can consider it to be $\hbar \partial / i \partial N$ when it is applied to a function of N. Then we have

$$e^{\pm i\Theta/\hbar} = e^{\pm \partial/\partial N} = 1 \pm \frac{\partial}{\partial N} + \frac{1}{2!} \frac{\partial^2}{\partial N^2} \pm \frac{1}{3!} \frac{\partial^3}{\partial N^3} + \ldots.$$

By Taylor's theorem

$$\psi(N) \pm \psi'(N) + \frac{1}{2!} \psi''(N) \pm \frac{1}{3!} \psi'''(N) + \ldots = \psi(N \pm 1),$$

therefore

$$e^{\pm i\Theta/\hbar} \psi(N) = \psi(N \pm 1). \quad \text{Q.E.D.}$$

Since the newly discovered theory has now been proved to be correct for bosons, the use of heuristic theory notwithstanding, there is no reason for us to hesitate to use it. We can proceed with it with confidence. Therefore we shall transform the Hamiltonian (6-9'), the equations of motion (6-10'), and the commutation relation (6-14) together with the relation between A_n and N (6-12') so we can use them for further discussions. Remembering (6-2), we define the "wave function" $\Psi(\mathbf{x})$ and its "conjugate momentum function" $\Pi(\mathbf{x})$, which are q-numbers, by

$$\Psi(\mathbf{x}) = \sum_n A_n \phi_n(\mathbf{x}) \quad (6\text{-}2')$$

$$\Pi(\mathbf{x}) = \sum_n \Pi_n \phi_n^*(\mathbf{x}).$$

Then we obtain from (6-14) the canonical commutation relations between $\Psi(\mathbf{x})$ and $\Pi(\mathbf{x})$

$$\Psi(\mathbf{x})\Pi(\mathbf{x}') - \Pi(\mathbf{x}')\Psi(\mathbf{x}) = i\hbar\,\delta(\mathbf{x} - \mathbf{x}')$$
$$\Psi(\mathbf{x})\Psi(\mathbf{x}') - \Psi(\mathbf{x}')\Psi(\mathbf{x}) = 0 \qquad (6\text{-}14')$$
$$\Pi(\mathbf{x})\Pi(\mathbf{x}') - \Pi(\mathbf{x}')\Pi(\mathbf{x}) = 0,$$

and we can derive the equation of motion for Ψ as

$$\left[H(\mathbf{x}, \mathbf{p}) + \frac{\hbar}{i}\frac{\partial}{\partial t} \right]\Psi(\mathbf{x}, t) = 0. \qquad (6\text{-}1')$$

Furthermore, since we see

$$\Pi(\mathbf{x}) = i\hbar\Psi^{\dagger}(\mathbf{x}), \qquad (6\text{-}8'')$$

the equation of motion of $\Pi(\mathbf{x})$ is nothing but the complex conjugate of (6-1'). (Since Ψ is a q-number, we use $\dagger$ instead of $*$ to express the complex conjugate.) Furthermore, the Hamiltonian (6-9') becomes

$$\overline{H} = \frac{1}{i\hbar}\int \Pi(\mathbf{x})H(\mathbf{x}, \mathbf{p})\Psi(\mathbf{x})dv = \int \Psi^{\dagger}(\mathbf{x})H(\mathbf{x}, \mathbf{p})\Psi(\mathbf{x})dv, \quad (6\text{-}5')$$

and the relation (6-12') becomes

$$N = \int \Psi^{\dagger}(\mathbf{x}, t)\Psi(\mathbf{x}, t)dv. \qquad (6\text{-}12'')$$

This N commutes with $\overline{H}$, and therefore we see it is independent of time.

Now if we look at (6-1') just obtained, we see it has a similar form to that of the probability amplitude of one particle, and if we look at the Hamiltonian (6-5'), then we see it has the same form as equation (6-5) for the expectation value of the energy for one particle. We should not forget, however, that although they are similar in form, their meanings are entirely different. Namely, (6-1') and (6-5') are related to the observables of "a mechanical system composed of N bosons," and both are relations between q-numbers, whereas (6-1) and (6-5) are related to the probability amplitude and the expectation value of one particle. We should note here that neither (6-1'), (6-5'), nor the commutation relation (6-14') contains the boson number N at all. We therefore can consider those relations as the fundamental formulas for "the mechanical system with an arbitrary number of bosons." Because of this, the wave function $\Psi(\mathbf{x})$, unlike the wave ψ in the coordinate space, is always a function of $\mathbf{x}$ in three-dimensional space, which is independent of the number of particles. We therefore can regard the wave function Ψ, which is our q-number, as the wave which actually exists in the three-dimensional space in which we live. Where does the number of particles appear then? The answer is in (6-12"). Namely, the number of particles appears related to the amplitude of Ψ.

We now know that the wave field which has the Hamiltonian (6-5') and is quantized by the commutation relation (6-14') is equivalent to the system of bosons with the Hamiltonian (6-16). Indeed, if we consider Ψ as a q-number and define A_n by (6-2'), then the eigenvalues of the q-number N_n, defined by

$$N_n = A_n^\dagger A_n, \tag{6-18}$$

are

$$\text{eigenvalues of } N_n = 0, 1, 2, \ldots, \tag{6-18'}$$

and therefore we see clearly the particle nature of the field. Furthermore, if we adopt Schrödinger's idea that $e\psi^*(\mathbf{x})\psi(\mathbf{x})$ is the charge density which actually exists in space and define the observable

$$\rho(\mathbf{x}) = e\Psi(\mathbf{x})^\dagger\Psi(\mathbf{x}), \tag{6-19}$$

(so far we have been using as electric charge $-e$, but today I shall use e for the charge of a particle regardless of the sign) then we see that the eigenvalues of its integral over arbitrary volume V

$$\rho_V = \int_V \rho(\mathbf{x})d\mathbf{x} \tag{6-19'}$$

are $0, 1e, 2e, 3e, \ldots$. This shows that our mechanical system has the property of an assembly of particles with electric charge e. We should note here that corresponding to the spreading of Schrödinger's $e\psi^*(\mathbf{x})\psi(\mathbf{x})$ in time, the expectation value of $e\Psi^\dagger(\mathbf{x})\Psi(\mathbf{x})$ is also blurred with time, and therefore the expectation value of ρ_V also may take a nonintegral value and gradually approaches 0. However, the eigenvalue of ρ_V will never take values other than 0 or positive integers. Therefore, although the expectation values may blur, the particle nature of electric charge is always conserved.

In this way Schrödinger's unfulfilled wish, namely not to place the wave $\psi(\mathbf{x})$ in the configuration space but to welcome it to the three-dimensional space, has now been realized by using the quantized $\Psi(\mathbf{x})$ rather than $\psi(\mathbf{x})$. Thus guided by his heuristic method, Dirac discovered that the field equation satisfied by

$\Psi(\mathbf{x})$ has the same form as the equation satisfied by $\psi(\mathbf{x})$. You should never forget, however, that even though the form of the equations is the same, ψ is the probability amplitude of one particle and is therefore a c-number, whereas Ψ, which describes the wave field, is a q-number, and they are conceptually entirely different things. Furthermore, as I shall tell you later, the coincidence of the form of the equations is limited only to cases in which the interactions between particles are neglected; if there are interactions, ψ and Ψ are not only conceptually different, but the equations satisfied by them have mathematical properties which are essentially different from each other. It is often said, "by the second quantization of ψ we obtain Ψ," but this statement is wrong for this reason. In my opinion, just as there are Maxwell equations which are not quantized, it is better to consider that there exist some equations from the beginning satisfied by nonquantized Ψ, and those equations agree with the equations of ψ only when there is no interaction.

Anyhow, it was a great discovery to have found that the mechanical system with the Hamiltonian (6-5'), i.e., the system of the wave field with equation (6-1'), and the mechanical system with Hamiltonian (6-16), i.e., a particle system composed of N particles, give exactly the same answer if we quantize the former by (6-14') and adopt only the symmetric wave function for the latter. They are entirely equivalent. It is great because from this we can establish a Nishida-like theme[1] in quantum mechanics that says a wave is a particle and a particle is a wave without any contradiction and obtain its complete mathematical expression.

Dirac's heuristic study in the form I have considered so far does not work for a real ensemble in which there is an interaction between particles. This is because in the virtual ensemble, which he used as a starting point, particle interactions do not play any role. After all, an individual mechanical system in his virtual ensemble is a system of *one* particle. Therefore, there is no interaction in the system. Also, the ensemble is virtual, and we may consider, for example, that the first system exists on the first day only, the second on the second day only, the third on the third day only, etc. Therefore, it is completely meaningless to consider the interaction between particles in two different systems. Nevertheless, Jordan and Klein have shown that it is possible to describe a real ensemble of particles which are interacting with each other through a wave field $\Psi(\mathbf{x})$ which actually exists in three-dimensional space if the particles are bosons. According to them, while so far for $V(\mathbf{x})$ in the Hamiltonian

---

1. Kitaro Nishida (1870–1945) was a renowned Japanese philosopher and writer. He founded his own school of philosophy based on the precept that "pure experience" is the sole reality. His philosophy attempted in his later years to approximate the concept of *Mu* (nothingness) of Zen. Tomonaga's father, Sanjuro (1871–1951) was a contemporary professor with Nishida in the department of philosophy, Kyoto University.

$$H(\mathbf{x}, \mathbf{p}) = \frac{1}{2m}\mathbf{p}^2 + V(\mathbf{x}) \qquad (6\text{-}20)$$

of field equation (6-1') we considered only potential energy due to the external field, but if there is an interaction between particles, for example, Coulomb repulsion, then we must add to the potential energy $V(\mathbf{x})$ of the external field a potential energy

$$V_{\text{wave}}(\mathbf{x}) = e \int \frac{e\Psi^\dagger(\mathbf{x}')\Psi(\mathbf{x}')}{|\mathbf{x} - \mathbf{x}'|}\, dv', \qquad (6\text{-}21)$$

which is caused by the charge density $e\Psi^\dagger(\mathbf{x})\Psi(\mathbf{x})$ of the wave field itself, and use

$$V'(\mathbf{x}) = V(\mathbf{x}) + V_{\text{wave}}(\mathbf{x}).$$

Namely, as a Hamiltonian in (6-1') we use

$$H(\mathbf{x}, \mathbf{p}) = \frac{1}{2m}\mathbf{p}^2 + V(\mathbf{x}) + V_{\text{wave}}(\mathbf{x}) \qquad (6\text{-}20')$$

and the equation

$$\left[\frac{1}{2m}\mathbf{p}^2 + V(\mathbf{x}) + e \int \frac{e\Psi^\dagger(\mathbf{x}')\Psi(\mathbf{x}')}{|\mathbf{x} - \mathbf{x}'|}\, dv' + \frac{\hbar}{i}\frac{\partial}{\partial t}\right]\Psi(\mathbf{x}, t) = 0 \quad (6\text{-}22)$$

instead of (6-1'). Jordan and Klein discovered that if we do this, this theory and the quantum theory of a boson system with the Hamiltonian

$$H = \sum_{v=1}^{N}\left[\frac{1}{2m}\mathbf{p}_n^2 + V(\mathbf{x})\right] + \sum_{v>v'}^{N}\frac{e^2}{|\mathbf{x}_v - \mathbf{x}_{v'}|} \qquad (6\text{-}23)$$

give results that agree in all respects. It is worth noting that in the second summation of (6-23) the term with $v = v'$ is excluded. As a result of this, for the case of one particle, the term involving $e^2/|\mathbf{x}_v - \mathbf{x}_{v'}|$ does not appear in (6-23) even if we use (6-22), in which $V_{\text{wave}}(\mathbf{x})$ is included.

From this work of Jordan and Klein, the difference between Ψ and ψ becomes even clearer. The field equation (6-22) for Ψ has an entirely different form from (6-1), which gives the probability amplitude of one particle, i.e.,

$$\left[\frac{1}{2m}\mathbf{p}^2 + V(\mathbf{x}) + \frac{\hbar}{i}\frac{\partial}{\partial t}\right]\psi(\mathbf{x}, t) = 0. \qquad (6\text{-}22')$$

Moreover, this difference is fundamental. The reason is that, according to the transformation theory, the probability amplitude must satisfy the principle of superposition, and therefore the equation satisfied by ψ should always be in linear form, but (6-22) is not linear with respect to Ψ. For this reason even if we consider that Ψ is not a q-number, the field equation (6-22) is of such a nature that it can never be considered as the equation for ψ. We now see clearly that it is entirely incorrect to say, "Ψ is obtained by the second quantization of ψ."

Equation (6-22), although not derivable by heuristic reasoning alone, is a field equation which takes into account the interaction of particles correctly, and if Ψ is not quantized, it should correspond to the classical Maxwell equation. Do you ask, Why then is there an $\hbar$ on the left-hand side of (6-22)? That is a good question. It is the sort of question that a happy-go-lucky person cannot ask. Since you noticed this very good point, try to answer it yourself. (Hint: Replace in the argument $m \rightarrow \hbar\hat{m}$, $V \rightarrow \hbar\hat{v}$, $e \rightarrow \hbar\hat{e}$, and $\Psi \rightarrow \hat{\Psi}/\sqrt{\hbar}$. Consider the meaning of these replacements.)

I have already talked for an unintentionally long time about Dirac's acrobatics, but all he wanted to show was that a system of many bosons and the wave field in three-dimensional space are equivalent in quantum mechanics. He tried to apply this conclusion to photons to discuss quantum-mechanically the emission, absorption, and scattering of photons by atoms. You might think that in order to apply the argument used so far directly to photons, we need the equation of probability amplitude of one photon. However, the photon is a relativistic particle, and we cannot use (6-1) for it. Many people at that time attempted without success to find a probability amplitude for one photon. (It was shown later that in the relativistic theory, not only for the case of photons, but in general, the probability amplitude in the x, y, z space does not exist.) However, as has now become clear after this long talk, it is the field equation that has to be quantized and not that of the probability amplitude. Therefore, even if we do not know the equation of probability amplitude, it suffices just to quantize the field equation, if it is known. Dirac knew this right from the beginning without needing this long exposition. He showed that by extending Debye's idea the correct answer for the absorption and emission of photons and the scattering of photons by atoms is obtained by quantizing the radiation field. Following this work of Dirac's, Fermi as well as Heisenberg and Pauli formalized more completely the interaction between an atom and the electromagnetic field by considering the electron and the quantized Maxwell field simultaneously. In particular, in their massive paper titled "On the Quantum Mechanics of the Wave Field" (1929), Heisenberg and Pauli, unlike Dirac and Fermi, treated the problem by considering not only the electromagnetic field but also the electron itself as a quantized field. (In this case, however, quantization cannot be done by (6-14'), for if so, the electron would be a boson. Then how to quantize an electron? I shall talk about this shortly.) In other words, in this paper they consider the Dirac equation not as the equation for the probability amplitude of an electron but as the relativistic field equation for the electron.

If we accept these premises, then probably it is all right to adopt as another possible relativistic field equation the Klein-Gordon equation, which was rejected by Dirac as an unsuitable equation for probability amplitude. Therefore, probably it is not that surprising that on one clear day[2] it occurred to Pauli to consider the Klein-Gordon equation and to quantize it with the Maxwell equations in a similar vein as the Heisenberg-Pauli theory.

Therefore he wrote a paper in 1934, helped by his assistant Victor Weisskopf. The paper's theme, "Nature has no reason to reject particle of spin zero," is the subtitle of my lecture today. Now we really head for the crux of the matter, but before doing that, let me finish the aforementioned problem, namely how to quantize the electron field.

Thus far I have told you that the particle which appears from the quantization of the wave field is a boson, but we would also like to know if we can treat a fermion like the electron by the method of field quantization. It was the work by Jordan and Wigner which appeared in 1928, one year before the work of Heisenberg and Pauli, which answered this question. Their answer was affirmative. However, we cannot, of course, use the commutation relation (6-14'), but instead we must use the relation in which the minus sign on the left-hand side of (6-14') is replaced by a plus sign.

$$\Psi(\mathbf{x})\Pi(\mathbf{x}') + \Pi(\mathbf{x}')\Psi(\mathbf{x}) = i\hbar\delta(\mathbf{x} - \mathbf{x}')$$
$$\Psi(\mathbf{x})\Psi(\mathbf{x}') + \Psi(\mathbf{x}')\Psi(\mathbf{x}) = 0 \qquad (6\text{-}14')_+$$
$$\Pi(\mathbf{x})\Pi(\mathbf{x}') + \Pi(\mathbf{x}')\Pi(\mathbf{x}) = 0.$$

They discovered these relations, which are called *anticommutation relations*. When these relations exist, the eigenvalues of the observable N_n defined by (6-18) are

$$\text{eigenvalues of } N_n = 0, 1, \qquad (6\text{-}18')_+$$

and therefore either 0 or 1 of this particle can exist in the state n, and thus Pauli's exclusion principle is clearly satisfied. Furthermore, it is possible to prove that the wave field thus quantized is completely equivalent in all respects to the solution in which only the antisymmetric wave function of the particle system with the Hamiltonian (6-16) or (6-23) is adopted, i.e., a system of fermions. Therefore the theme which I told you earlier à la Nishida's philosophy works for both bosons and fermions.

---

2. Tomonaga uses the phrase *Un bel dì* from the aria of *Madame Butterfly*.

Now we will go into the story of Pauli and Weisskopf, which is the main theme of this lecture. Since I had to lay the groundwork by delving into the huge problem of wave equals particle and particle equals wave, I have nearly run out of time. However, because of this long preparation you probably have understood the background for Pauli and Weisskopf's work, so let me just give you their results.

As I told you earlier, Pauli and Weisskopf tried to quantize the Klein-Gordon equation together with the Maxwell equations using (6-14'). They were able to do so without any inconsistencies, and from the Klein-Gordon field they obtained bosons of mass m and spin 0, and very interestingly they found that the electric charge of this boson can take $+e$ or $-e$, i.e., be either positive or negative. Moreover, not only was this possible, but they also found through calculation that when there exists a photon whose $h\nu$ is larger than $2mc^2$, then its absorption can create a pair of particles with $+e$ and $-e$, and conversely, if there is a $+e/-e$ pair, then they can be annihilated by emitting $h\nu > 2mc^2$.

Furthermore, Pauli and Weisskopf objected to regarding the Dirac equation as the equation of relativistic probability amplitude of one electron although it is first order. According to them, the Dirac equation is also the relativistic *field* equation for the electron and it cannot be considered to be an equation of probability amplitude in x, y, z space. They insisted that a concept like "the probability of a particle to be at **x** in space" is meaningless for relativistic particles—be they electrons, photons, or Klein-Gordon particles—and therefore it is meaningless to interpret $\psi(\mathbf{x})$ as the probability amplitude.

One basis for this last declaration is that, if we consider the Dirac equation to be the equation of probability amplitude of an electron, then there appears the peculiar state in which the electron has a negative energy, and the positive-energy electron will fall into the negative-energy state by emitting energy as a result of interaction with the electromagnetic field. If this can happen, many peculiar phenomena which contradict reality will follow. In order to cope with this difficulty, around 1930 Dirac introduced the hypothesis that the vacuum is the state in which all the negative-energy levels are occupied by electrons. Then because of the Pauli principle, the positive-energy electrons cannot fall into negative energy. In Pauli's opinion, if all the negative-energy levels are filled by so many electrons, there will be an infinite number of electrons, and that is already way beyond the one-body problem.

For these reasons Pauli deemed that, just as in the Heisenberg-Pauli paper, we should treat the Dirac equation as a field equation rather than as an equation for probability amplitude. Dirac, apparently, did not care for Pauli taking Dirac's equation as the field equation and proposed in his new theoretical formulation of the many-electron problem published in 1932, which later came to be called the many-time theory, to use the probability amplitude ψ in the coordinate system [however, in order to make ψ relativistic we extend the concept of coordinates in the wave function by assigning a different time for each electron, hence $\psi(\mathbf{x}_1, t_1, \mathbf{x}_2, t_2, \mathbf{x}_3, t_3, \ldots)$].

Now the other side of the story is that Dirac himself was starting to anticipate from his hypothesis of filling the negative-energy levels that in addition to the usual electron, i.e., an electron with negative charge, there exists an electron with positive charge, just as bosons of $+e$ and $-e$ appear from quantizing the Klein-Gordon field. His idea was that if the filling of negative-energy levels is incomplete and an electron is missing from some negative-energy level, then that "hole" has a positive energy and behaves as if it has a positive charge. (Dirac's idea is that lack of negative must mean positive.) Therefore, a hole would look like an electron with positive charge. (Initially, Dirac thought that this hole was a proton. However, Oppenheimer raised the criticism that according to this idea, the hole would be immediately filled by a nearby electron and the hydrogen atom could not exist stably. Furthermore, it was pointed out by the mathematician Hermann Weyl that this hole should behave as if it had the same mass as an electron.)

According to this idea, if there is a photon with $h\nu > 2mc^2$, then an electron in a negative-energy level is excited to a positive-energy level, absorbing energy, and as a result an electron with positive energy and a hole in a negative level—in other words, a positive-energy, positive-charge electron—are created. By the way, this positively charged electron was discovered experimentally in 1932 by Carl Anderson and was named *positron*.

I told you that both positively and negatively charged particles can appear when the Klein-Gordon equation is quantized. However, Dirac already pointed out in his 1928 paper, which I talked about in lecture 3, that even without quantization the equation has one solution which behaves like a particle with negative charge and one which behaves like a particle with positive charge. (I shall discuss this further in the next lecture.) He gave this as one of the reasons for insisting that the electron cannot be described by this equation. (At that time only electrons with negative charge were believed to exist.)

However, after the discovery of the positron, not only did this reason become groundless but it also became apparent that both bosons related to the Klein-Gordon equation and fermions related to the Dirac equation have a very similar property in that they have positive and negative charge, and from these resemblances it became quite natural to believe that both the Klein-Gordon equation and the Dirac equation have equally good raison d'être. For these reasons Pauli confidently pushed for the resurrection of the Klein-Gordon field. Furthermore, if you quantize the Dirac equation in the manner of Heisenberg and Pauli, then it is possible with some ingenuity to incorporate positrons and pair creation into the theory without any artificial postulates such as filling the negative-energy levels and holes. Because of these points, Pauli thought it much more natural to regard the Dirac equation as the field equation rather than as the equation of probability amplitude as Dirac preferred.

Thus Pauli and Weisskopf reinstated the Klein-Gordon equation, which Dirac had rejected. It seems Pauli was quite euphoric about this work, and a part of the paper sounds as if he is teasing Dirac by using exactly the phrase which Dirac had used in his other paper. Freely translated, it reads as follows.

"The most interesting part of our theory is that the energy is always positive automatically (namely, without using a superfluous hypothesis such as hole theory). After witnessing that the relativistic scalar theory (the theory of the Klein-Gordon field) can be constructed without any such hypothesis, one might wonder *why nature had made no use* of the possibility from the theory that there exist spin-zero bosons with charge $\pm e$ and the possibility that they can be created from $h\nu$ or annihilated, emitting $h\nu$." The italicized phrase is taken directly from the paper in which Dirac predicted the existence of a certain particle (magnetic monopole). It seems to me that the tone of this sentence and that of Dirac's sentence (which I quoted in lecture 3) in the beginning of his paper on the Dirac equation are nearly identical.

This sentence sounds as if Pauli wanted to say that there exist charged particles with spin 0, but when we read on we find on the contrary that he sounds as if he is looking for some reason why such a particle cannot be found. However, when Pauli was writing this paper in 1934, the new idea of the meson was already taking shape in Yukawa's mind, and the relativistic scalar theory (more accurately, pseudoscalar theory) would play a big role in the theory of mesons. Therefore, nature indeed did use this possibility in the appearance of π mesons.

All the same, Pauli and Weisskopf's work and Yukawa's idea of the meson appeared one after the other with incredible timing. The history of physics can sometimes be very theatrical.

The Quantity Which Is Neither Vector nor Tensor

The Discovery of the Spinor Family and the Astonishment of Physicists

In the previous lecture we proceeded up to 1934, but as I told you, we were led up to Pauli and Weisskopf's work by a chain of discoveries starting in 1927. Among them one of the central issues was the discovery that we can regard the matter wave as a wave in three-dimensional space. There is one more interesting development, however, between 1927 and 1930, and since it relates to spin, we must delve into it. Therefore, let me return once again to 1927–1928.

In lecture 3 I discussed the attempt to incorporate spin into the framework of quantum mechanics, and how from this attempt the Pauli equation was found in 1927 and the Dirac equation in 1928. In these equations two-component quantities and four-component quantities were used, and, for example, in Pauli's equation ψ had the form

$$\psi = \begin{pmatrix} \psi_1(x, y, z, t) \\ \psi_2(x, y, z, t) \end{pmatrix}. \tag{7-1}$$

We have written the components ψ_1 and ψ_2 as a column vector because this form is convenient for multiplication with the 2×2 spin matrices. The problem arising from this formalism is to find the transformation property of these multicomponent quantities.

Multicomponent quantities have frequently been used in physics since the old days. For example, in mechanics or electrodynamics vectors and tensors which have three and nine components, respectively, have been used. Especially in electrodynamics, the electric and magnetic fields are field quantities, and unlike the velocity or acceleration of a mass point in mechanics, they are point functions and therefore functions of x, y, z (and t). In that sense they are very similar to (7-1). In addition to these multicomponent quantities,

113

Figure 7.1 Paul Ehrenfest
(1880–1933), ca. 1901
[Photograph by Leningrad
Physico-Technical
Institute. Courtesy
AIP Emilio Segrè
Visual Archives]

single-component quantities such as mass, charge, and potential have also been used in physics and are called scalars.

The vectors and tensors which we have been discussing are in the three-dimensional world, but as you know, when relativity appeared, the four-dimensional Minkowski world in which time is ranked alongside the space variables became the arena of physicists. Accordingly, many quantities became four-dimensional vectors or four-dimensional tensors, and tensor algebra was found to be extremely powerful in relativity theory. However, because of the limitation of time, we shall not explore the four-dimensional world today.

Now the problem thrust upon us by the discovery of the Pauli equation was, To which of the categories of scalar, vector, or tensor in three-dimensional space does the two-component quantity (7-1) belong? In order to answer this question, it is necessary to remember the definitions of vector and tensor and why physics needed them.

There are a variety of ways to define vector and tensor; however, the most suitable for today's discussion is to define them as *covariant quantities* with respect to transformations of the space coordinates. If we use this concept of *covariance*, then we get a good hint as to the significance of vectors and tensors in physics and the properties of two-component quantities such as (7-1).

Therefore, let us begin by reviewing the definitions of vector and tensor. Let us take an arbitrary point P in three-dimensional space. We consider an arbitrary orthogonal coordinate system R and let the coordinates of P in this system be $x_j (j = 1, 2, 3)$. We then consider another coordinate system R' which is obtained by rotating the coordinate axes of system R around the origin. Let the coordinates of point P in the R' system be $x'_k (k = 1, 2, 3)$. Then you know there are relations

$$x'_k = \sum_{j=1}^{3} A_{kj} x_j \qquad k = 1, 2, 3 \qquad (7\text{-}2)$$

between x_j and x'_k. Here, the nine coefficients A_{kj} are the direction cosines between the coordinate axes of the R system and those of the R' system, and therefore the matrix composed of A_{kj}

$$A = \begin{pmatrix} A_{11} & A_{12} & A_{13} \\ A_{21} & A_{22} & A_{23} \\ A_{31} & A_{32} & A_{33} \end{pmatrix} \qquad (7\text{-}3)$$

is orthogonal. Therefore by solving (7-2) we get

$$x_j = \sum_{k=1}^{3} x'_k A_{kj} \qquad j = 1, 2, 3. \qquad (7\text{-}2')$$

Furthermore, since the R' system is obtained from the R system by a rotation, if the R system is a right-handed system, then the R' system is, too, and in such a case

$$\det A = +1. \qquad (7\text{-}4)$$

As I told you, A_{kj} are direction cosines, and therefore if we consider orthogonal matrices with $\det A = \pm 1$, then the rotation from the R system to the R' system is determined uniquely and vice versa. Therefore, there is a one-to-one correspondence between the rotation of coordinate systems and matrices A, and we can call that rotation "rotation A."

Next, if we go from the R system to the R' system by rotation A and then go from the R' system to the R'' system by rotation B, then there is a relation

$$x_\ell'' = \sum_{j=1}^{3} C_{\ell j} x_j, \qquad \ell = 1, 2, 3, \tag{7-5}$$

between the coordinates of P in the R system (written x_j) and those in the R'' system (written x_ℓ''), and the matrix of $C_{\ell j}$

$$C = \begin{pmatrix} C_{11} & C_{12} & C_{13} \\ C_{21} & C_{22} & C_{23} \\ C_{31} & C_{32} & C_{33} \end{pmatrix} \tag{7-6}$$

is the product of the matrices A and B,

$$C = B \cdot A. \tag{7-7}$$

We describe this fact by saying that "performing rotation A and then rotation B on the result is equal to performing rotation $B \cdot A$."

I have given a long introduction, but now let us define vectors. Let us consider a three-component quantity

$$(a_1, a_2, a_3). \tag{7-8}$$

When the coordinate system R is rotated to R' by A and if the components a_1, a_2, a_3 transform in the same way as (7-2), that is

$$a_k' = \sum_{j=1}^{3} A_{kj} a_j, \qquad k = 1, 2, 3, \tag{7-9}$$

then we call the three-component quantity (7-8) a vector. This is the definition of vector. From this definition it also holds that

$$a_j = \sum_{k=1}^{3} a_k' A_{kj}. \tag{7-9'}$$

According to this definition, the three-component quantity (x_1, x_2, x_3) which determines the position of a mass point is obviously a vector.

Now tensors are defined as follows. Let there be a nine-component quantity

$$\begin{pmatrix} a_{11} & a_{12} & a_{13} \\ a_{21} & a_{22} & a_{23} \\ a_{31} & a_{32} & a_{33} \end{pmatrix}. \tag{7-10}$$

(Although I arranged the a's in square form, I remind you this is not a matrix.) Since it is cumbersome to write the whole square each time, we shall write

$$(a_{jk}) \qquad j = 1, 2, 3; \quad k = 1, 2, 3 \tag{7-11}$$

instead of (7-10). Sometimes we will omit $j = 1, 2, 3; k = 1, 2, 3$.

When we have such a nine-component quantity whose components a_{jk} are transformed according to

$$a'_{\ell m} = \sum_{j=1}^{3} \sum_{k=1}^{3} A_{\ell j} A_{mk} a_{jk} \tag{7-12}$$

when the coordinate system is transformed from R to R' by the rotation A, then we call this (a_{jk}) a second-rank tensor. The coefficients $A_{\ell j} A_{mk}$ are written

$$A_{\ell j} A_{mk} = A_{\ell m, jk}, \tag{7-13}$$

and we define the 9×9 matrix

$$A^{(2)} \equiv (A_{\ell m, jk}) \qquad \ell, m = 1, 2, 3; \quad j, k = 1, 2, 3. \tag{7-14}$$

(Here in the parentheses are nine rows corresponding to $\ell, m = 1, 2, 3$ and nine columns corresponding to $j, k = 1, 2, 3$.) Using this matrix, we can write (7-12) in the form

$$a'_{\ell m} = \sum_{j,k=1}^{3} A_{\ell m, jk} a_{jk}; \tag{7-15}$$

we can also write

$$a_{jk} = \sum_{\ell, m=1}^{3} a'_{\ell m} A_{\ell m, jk}. \tag{7-15'}$$

These are the relations corresponding to (7-9) and (7-9') for vectors.

In mathematics we define

$$A_{\ell j} B_{mk} = C_{\ell m, jk} \tag{7-16}$$

from the two matrices $A = (A_{\ell j})$ and $B = (B_{mk})$, and the matrix composed of these

$$C = (C_{\ell m, jk}) \tag{7-17}$$

we call the "direct product of A and B" and write

$$C = A \times B. \tag{7-16'}$$

If I use this terminology, the $A^{(2)}$ of (7-14) is said to be the direct product of A and A, namely the square

$$A^{(2)} = A \times A. \tag{7-18}$$

Therefore we may say, "A second-rank tensor is a quantity whose components are transformed by $A^{(2)} = A \times A$, corresponding to the rotation A of the coordinate system."

In this way the definitions of third-rank tensor, fourth-rank tensor, etc., are automatically clear. Namely, a third-rank tensor is a quantity with 3^3 components that are transformed by

$$A^{(3)} = A \times A \times A.$$

According to this definition, we can consider a vector as a quantity that transforms as $A^{(1)} = A$, corresponding to the rotation A of the coordinate system, and in this sense we sometimes call a vector a first-rank tensor. We could also say a scalar is a quantity which is transformed by $A^{(0)} = 1$ and call it a zeroth-rank tensor. According to this definition, it is not sufficient to say that a scalar is a quantity with one component, but it must be a quantity which does not change its value in any coordinate system.

From these definitions we can immediately derive the following. First, regardless of the rank of a tensor, the transformation of the components of the tensor and the rotation of the coordinates always have a one-to-one correspondence. Therefore, if there are rotations of the coordinate system A, B, and C, then to each of them corresponds a transformation matrix of tensor components $A^{(n)}$, $B^{(n)}$, $C^{(n)}$. Furthermore, we can prove that if $C = B \cdot A$, then there exists a relation between matrices $A^{(n)}$, $B^{(n)}$, $C^{(n)}$

$$C^{(n)} = B^{(n)} \cdot A^{(n)}.$$

For these reasons, not only is there a one-to-one relation between the rotation of coordinate systems and the transformation matrix of tensor components of arbitrary rank, but there also is a correspondence *product to product*. The existence of this type of correspondence is expressed by saying that the transformation of tensor components and the rotation of coordinate systems *are covariant*. In this sense we call vectors and tensors *covariant quantities*. Since you probably know group theory, I mention in passing that $A^{(n)}$ is the 3^n-dimensional representation of the rotation group, and a tensor of nth rank is a quantity with 3^n components which are transformed by $A^{(n)}$.

We shall not now go into the variety of operations which is used in the theory of vectors and tensors. Among these operations are the addition of vectors, the product of vectors, and the contraction of tensors. All of them make vectors or scalars from vectors, or vectors from tensors or vectors, or scalars from tensors, etc., and they all make covariant quantities from covariant quantities. Operations that transform covariant quantities into anything other than covariant quantities are useless in physics.

So why are only covariant quantities useful in physics? Let me explain using an example. I told you the coordinate (x_1, x_2, x_3) which represents the position of a mass point is indeed a vector. Then the three-component quantity $(\dot{x}_1, \dot{x}_2, \dot{x}_3)$, which is called velocity, and the three-component quantity $(\ddot{x}_1, \ddot{x}_2, \ddot{x}_3)$, which is called acceleration, are surely vectors as we see if we differentiate both sides of (7-2) with respect to time. Now we know that the three-component quantities force and acceleration are related through the simultaneous equations

$$m\ddot{x}_1 = f_1 \qquad m\ddot{x}_2 = f_2 \qquad m\ddot{x}_3 = f_3. \qquad (7\text{-}19)$$

Here m is the mass, and it is a constant specific to the mass point under consideration. Now we know from Newton's laws there is a fundamental postulate that the physical space is isotropic, and the laws of physics should not change regardless of the orientation of the orthogonal coordinate system used. Therefore, the simultaneous equations (7-19) should hold in the same form for any coordinate system. As I told you, $(\ddot{x}_1, \ddot{x}_2, \ddot{x}_3)$ on the left-hand side of (7-19) is a vector, and it transforms according to (7-9) when we move from the R system to the R' system. Then since m is a constant specific to the mass point (accordingly, a scalar), the right-hand side (f_1, f_2, f_3) must also transform according to (7-9). For if this were not the case, the system of simultaneous equations (7-19) would take a different form in the R' system. When the simultaneous equations between the components of physical quantities take the same form in any coordinate system as in this example, we say, "the equations are covariant" (or we may also say "the equations are invariant").

Perhaps this example is too simple, but the situation is the same even for more complicated laws of physics. Even when complicated and abstract physical quantities appear in a law, in order for the physical law to be confirmed experimentally there must be something like force acting on the needle of a meter somewhere in the equations expressing the law. Thus the complicated relations of physical quantities can be untangled strand by strand, and eventually only relations among covariant quantities appear in the physical laws. For this reason we can say that all equations in physics must be in covariant form.

Now let me here add two snake's legs (redundant comments) because they relate to our discussion later. One concerns quantities such as potential or electromagnetic field, which are, respectively, scalar or vector functions of x_1, x_2, x_3. If we take a simple case, e.g., potential, the question is, If it is

$$\phi(x_1, x_2, x_3) \tag{7-20}$$

in the R system, what would it be in the R' system? The answer is that the potential in the R' system is $\overline{\phi}(x'_1, x'_2, x'_3)$, defined by

$$\overline{\phi}(x'_1, x'_2, x'_3) \equiv \phi\left(\sum_k x'_k A_{k1}, \sum_k x'_k A_{k2}, \sum_k x'_k A_{k3}\right). \tag{7-20'}$$

This is because since the potential is a scalar, its value at point P must be the same in whichever system we use. Indeed if we use (7-2') in the right-hand side of (7-20'), then

$$\overline{\phi}(x'_1, x'_2, x'_3) = \phi(x_1, x_2, x_3). \tag{7-21}$$

We must note that although the *values* of $\overline{\phi}$ on the left-hand side and of ϕ on the right-hand side are equal, the *functional forms* of $\overline{\phi}$ and ϕ are different.

In a similar way a vector in the R system

$$\left(E_j(x_1, x_2, x_3)\right), \qquad j = 1, 2, 3 \tag{7-22}$$

is transformed in the R' system and becomes

$$\left(\sum_k A_{kj}\overline{E}_j(x'_1, x'_2, x'_3)\right), \qquad k = 1, 2, 3. \tag{7-22'}$$

Here the function $\overline{E}_j$ is defined by

$$\overline{E}_j(x'_1, x'_2, x'_3) \equiv E_j\left(\sum_k x'_k A_{k1}, \sum_k x'_k A_{k2}, \sum_k x'_k A_{k3}\right). \tag{7-23}$$

We can regard the transformation (7-22') as if we transform the components

$E_j(x_1, x_2, x_3)$ with (7-20'), regarding them as mutually independent scalars, and then use (7-9) to remind ourselves that they are components of a vector.

In the discussion above, the independent variables for a function without a bar are always x_1, x_2, x_3, and those for a function with a bar are always x_1', x_2', x_3'. Therefore, from now on we do not write the variables when we use the unbarred and barred functions.

I would also like to point out that when a scalar or a vector is a function of x_1, x_2, x_3, then certain kinds of differential operators can be regarded as scalars or tensors. For example, the well known $\nabla \equiv (\partial/\partial x_1, \partial/\partial x_2, \partial/\partial x_3)$ can be regarded as a vector. The meaning of that is that if we "multiply" them to a scalar ϕ

$$\left(\frac{\partial \phi}{\partial x_1}, \frac{\partial \phi}{\partial x_2}, \frac{\partial \phi}{\partial x_3} \right), \tag{7-24}$$

then the relations

$$\frac{\partial \bar{\phi}}{\partial x_k'} = \sum_j A_{kj} \frac{\partial \phi}{\partial x_j}, \qquad k = 1, 2, 3 \tag{7-24'}$$

hold. Namely, in the sense that if we operate ∇ on a scalar, we obtain a vector, and therefore we consider ∇ itself as a vector. For the same reason the familiar $\Delta \equiv (\nabla \cdot \nabla)$ can be regarded as a scalar because if we operate on ϕ, we obtain

$$\left(\frac{\partial^2}{\partial x_1'^2} + \frac{\partial^2}{\partial x_2'^2} + \frac{\partial^2}{\partial x_3'^2} \right) \bar{\phi} = \left(\frac{\partial^2}{\partial x_1^2} + \frac{\partial^2}{\partial x_2^2} + \frac{\partial^2}{\partial x_3^2} \right) \phi. \tag{7-25}$$

All these things were known by 1910–1911. We now jump to 1927–1928 and discuss the Pauli equation.

First let us remember the Pauli theory, which we described in lecture 3. I told you that Pauli used the matrices

$$\sigma_1 = \begin{pmatrix} 0 & 1 \\ 1 & 0 \end{pmatrix} \qquad \sigma_2 = \begin{pmatrix} 0 & -i \\ i & 0 \end{pmatrix} \qquad \sigma_3 = \begin{pmatrix} 1 & 0 \\ 0 & -1 \end{pmatrix} \tag{7-26}$$

to formulate his theory. He assumed that the components of spin angular momentum measured in units of $\hbar$ are $\sigma_1/2, \sigma_2/2, \sigma_3/2$, and correspondingly the components of the electron spin magnetic moment measured in units of the Bohr magneton are

$$\mu_j = -\sigma_j \qquad j = 1, 2, 3. \tag{7-27}$$

He derived

$$\left\{ H_0 + H_S + \frac{\hbar}{i} \frac{\partial}{\partial t} \right\} \psi = 0, \tag{7-28}$$

as the Schrödinger equation satisfied by the two-component quantity $\psi = \binom{\psi_1}{\psi_2}$ from (7-1). Here the Hamiltonians H_0 and H_S are

$$H_0 = \frac{1}{2m} \sum_k p_k^2 - \frac{Ze^2}{r} + \frac{e\hbar}{2mc} \sum_k H_k \ell_k \tag{7-29}_0$$

$$H_S = \frac{e\hbar}{2mc} \sum_k H_k \sigma_k + \frac{1}{2} \frac{e\hbar}{2mc} \sum_k \overset{\circ}{H}_k \sigma_k. \tag{7-29}_S$$

H_0 is the part which is independent of spin, and H_S is the part which depends on spin. $\mathbf{H} = (H_1, H_2, H_3)$ is the external magnetic field, $\boldsymbol{\ell} = (\ell_1, \ell_2, \ell_3)$ is the orbital angular momentum, and $\boldsymbol{\sigma}/2 = (\sigma_1/2, \sigma_2/2, \sigma_3/2)$ is the spin angular momentum. The third term of H_0 is the interaction energy between $\mathbf{H}$ and the orbital magnetic moment, and the first term in H_S is the interaction energy between $\mathbf{H}$ and the spin magnetic moment. $\overset{\circ}{\mathbf{H}} = (\overset{\circ}{H}_1, \overset{\circ}{H}_2, \overset{\circ}{H}_3)$ in the second term of H_S is the "internal magnetic field," which was discussed in lecture 3 (from Biot-Savart's law $\overset{\circ}{\mathbf{H}} = (1/mc) (Ze/r^3) \hbar \boldsymbol{\ell}$, and the term represents the interaction energy of $\overset{\circ}{\mathbf{H}}$ and the spin magnetic moment. The $1/2$ in this term is the Thomas factor, which we discussed in lecture 3. [Please remember (3-8), (3-10), and (3-10').]

Now let us examine how the Pauli equation (7-28) changes when the coordinate system is varied. In other words we consider the coordinate system in which (7-28) is written and try to transform the formula to the R' system. Let us first look at H_0. The three terms in the expression are all scalars. This part of the Hamiltonian, therefore, is transformed to

$$H_0' = \frac{1}{2m} \sum_k p_k'^2 - \frac{Ze^2}{r} + \frac{e\hbar}{2mc} \sum_k H_k' \ell_k' \tag{7-29'}_0$$

in the R' system. Here H_0 is not just a number, but it is an operator on the function $\psi(x_1, x_2, x_3)$; therefore, although it is a scalar, we do not have $H_0' = H_0$ but

$$H_0' \overline{\psi} = H_0 \psi, \tag{7-30}_0$$

analogous to (7-25). If this relation holds, then as far as the Hamiltonian H_0 is concerned, the form of the Pauli equation in the R' system is identical to that in the R system.

Now what happens to H_S? First we combine the two terms of H_S and write

$$H_S = \sum_k V_k \sigma_k. \qquad (7\text{-}29)_{\bar{S}}$$

Here $\mathbf{V} = (V_1, V_2, V_3)$ is a vector obtained from the combination of $\mathbf{H}$ and $\overset{\circ}{\mathbf{H}}$. Now let us transform the coordinate system from R to R'. Since H_S is a scalar product of $\mathbf{V}$ and $\boldsymbol{\sigma}$, H_S is a scalar, and therefore H_S' must have the "same form" as $(7\text{-}29)_{\bar{S}}$

$$H_S' = \sum_k V_k' \sigma_k'. \qquad (7\text{-}29')_{\bar{S}}$$

Here of course

$$\sigma_k' = \sum_j A_{kj} \sigma_j \qquad k = 1, 2, 3. \qquad (7\text{-}31)$$

Therefore, we also have

$$H_S' \,\overline{\psi} = H_S \psi. \qquad (7\text{-}30)_S$$

However, unlike the case of H_0, there is no guarantee that the equation for a two-component quantity is in covariant form because of $(7\text{-}30)_S$. The reason for this is that the definition of covariant form states that when the simultaneous equations of components are written, we obtain the same formula for any coordinate system. If $\sigma_1, \sigma_2, \sigma_3$ in H_S were ordinary numbers, then just as in H_0 there would be no problem, but in reality they are matrices, and while $\sigma_1, \sigma_2, \sigma_3$ are given by (7-26), $\sigma_1', \sigma_2', \sigma_3'$ are

$$\sigma_k' = \begin{pmatrix} A_{k3} & A_{k1} - i A_{k2} \\ A_{k1} + i A_{k2} & -A_{k3} \end{pmatrix} \qquad k = 1, 2, 3, \qquad (7\text{-}26')$$

as is evident from (7-31), and therefore the two components for $H_S \psi$ and the two components for $H_S' \,\overline{\psi}$ take completely different forms. In order for them to be of the same form, H_S' must be $H_S' = \sum_k V_k' \sigma_k$ rather than $\sum_k V_k' \sigma_k'$. [See (7-40) and (7-40'), which will be discussed later.]

How can we overcome this difficulty? We note the following point. So far in transforming from the R system to the R' system, we have been treating the two components ψ_1 and ψ_2 of ψ as mutually independent scalars. Is this not equivalent to simply applying only the transformation (7-22') and forgetting about the subsequent transformation (7-23) in transforming the vector (7-22)?

If this is the case, the two-component quantity to be used in R' should not be $\left(\begin{smallmatrix}\psi_1\\\psi_2\end{smallmatrix}\right)$ itself but something like $\left(\begin{smallmatrix}\overline{\psi}'_1\\\overline{\psi}'_2\end{smallmatrix}\right)$, which is obtained by a transformation of the form

$$\overline{\psi}'_\beta = \sum_{\alpha=1}^{2} U_{\beta\alpha}\overline{\psi}_\alpha \quad \beta = 1, 2. \tag{7-32}$$

Isn't this why the forms of the equations did not agree between the R system and the R' system?

Pauli has shown that indeed this is the case. He first noted that $\sigma_1, \sigma_2, \sigma_3$ of (7-26) and $\sigma'_1, \sigma'_2, \sigma'_3$ of (7-26') are related by using a unitary matrix,

$$\sigma'_1 = U^{-1}\sigma_1 U \qquad \sigma'_2 = U^{-1}\sigma_2 U \qquad \sigma'_3 = U^{-1}\sigma_3 U. \tag{7-33}$$

Perhaps you know this already, but the definition of a unitary matrix is as follows. When we interchange the rows and columns of a matrix $U = (U_{\beta a})$ and take the complex conjugate to obtain $U^\dagger = (U^*_{a\beta})$, then if there is the relation between U and $U^\dagger$

$$U^\dagger U = UU^\dagger = 1, \tag{7-34}$$

we call this matrix unitary. We shall not give the proof of (7-33) here, but it is a natural conclusion from a fundamental theorem used in transformation theory. I just point out the following, namely, that U is a matrix related to the rotation A from the R system to the R' system, but it is independent of x_1, x_2, x_3, t.

So we use (7-33) in the right-hand side of (7-29')$_\text{S}$. We then obtain

$$H'_S = U^{-1}\left(\sum_k V'_k\sigma_k\right) U.$$

From this we immediately obtain

$$H'_S \overline{\psi} = U^{-1}\left(\sum_k V'_k\sigma_k\right) U \overline{\psi},$$

and using (7-30)$_\text{S}$ for the left-hand side and applying U to both sides, we obtain

$$U H_S \psi = \left(\sum_k V'_k\sigma_k\right) U \overline{\psi}. \tag{7-35}$$

Therefore, if we introduce the two-component quantity $\overline{\psi}'$ by

$$\overline{\psi}' \equiv U\,\overline{\psi}, \tag{7-36}$$

then we obtain the relation

$$\left(\sum_k V'_k \sigma_k\right)\overline{\psi}' = U H_S \psi. \tag{7-37$_S$}$$

(Note that the left-hand side is not $\sum_k V'_k \sigma'_k$.)

So far we have been considering only H_S, but let us show that

$$H'_0\,\overline{\psi}' = U H_0 \psi. \tag{7-37$_0$}$$

As I told you earlier, U is independent of x_1, x_2, x_3, and therefore U commutes with both H_0 and H'_0. Therefore, we can rewrite the relation $U H'_0\,\overline{\psi} = U H_0 \psi$, which was obtained by applying U to (7-30)$_0$, as $H'_0 U\,\overline{\psi} = U H_0 \psi$. Then we can use for $U\,\overline{\psi}$ the left-hand side of (7-36), and we immediately get (7-30)$_0$. Therefore, (7-37)$_0$ is valid in addition to (7-37)$_S$, and adding them, we obtain

$$\left(H'_0 + \sum_k V'_k \sigma_k\right)\overline{\psi} = U(H_0 + H_S)\psi. \tag{7-37}$$

Now we are ready to discuss the Schrödinger equation. First, it is obvious that

$$\left(H'_0 + \sum_k V'_k \sigma_k + \frac{\hbar}{i}\frac{\partial}{\partial t}\right)\overline{\psi}' = \left(H'_0 + \sum_k V'_k \sigma_k\right)\overline{\psi}' + \frac{\hbar}{i}\frac{\partial}{\partial t}U\overline{\psi}. \tag{7-37'}$$

We write the first term on the right-hand side as $U(H_0 + H_S)\psi$ according to (7-37). Next, the second term can be written as $U(\hbar/i)\partial\overline{\psi}/\partial t)$ because U is independent of t, and then from (7-21) it becomes $U(\hbar/i)\partial\psi/\partial t$. Therefore, we derive for the right-hand side

$$\text{right-hand side of (7-37')} = U\left(H_0 + \sum_k V_k \sigma_k + \frac{\hbar}{i}\frac{\partial}{\partial t}\right)\psi.$$

Now since the Schrödinger equation is

$$\left(H_0 + \sum_k V_k \sigma_k + \frac{\hbar}{i} \frac{\partial}{\partial t} \right) \psi = 0 \tag{7-38}$$

for the R system, the right-hand side of (7-37') for R' is also 0, and we finally have

$$\left(H_0' + \sum_k V_k' \sigma_k + \frac{\hbar}{i} \frac{\partial}{\partial t} \right) \overline{\psi}' = 0. \tag{7-38'}$$

Now if we write everything down,

$$\psi = \begin{pmatrix} \psi_1 \\ \psi_2 \end{pmatrix} \qquad \overline{\psi}' = \begin{pmatrix} \overline{\psi}_1' \\ \overline{\psi}_2' \end{pmatrix}, \tag{7-39}$$

and therefore using the first of these formulas in (7-38) and using (7-26) for $\sigma_1, \sigma_2, \sigma_3$, we obtain the simultaneous equations

$$\begin{cases} H_0 \psi_1 + V_3 \psi_1 + (V_1 - i V_2)\psi_2 + \dfrac{\hbar}{i} \dfrac{\partial \psi_1}{\partial t} = 0 \\[2mm] H_0 \psi_2 + (V_1 + i V_2)\psi_1 - V_3 \psi_2 + \dfrac{\hbar}{i} \dfrac{\partial \psi_2}{\partial t} = 0 \end{cases} \tag{7-40}$$

as the Schrödinger equation for the system R. On the other hand, since in (7-38') the second term in parentheses is $\sum_k V_k' \sigma_k$ rather than $\sum_k V_k' \sigma_k'$, we get from the second formula of (7-39) precisely the same formulas

$$\begin{cases} H_0' \overline{\psi}_1' + V_3' \overline{\psi}_1' + (V_1' - i V_2') \overline{\psi}_2' + \dfrac{\hbar}{i} \dfrac{\partial \overline{\psi}_1'}{\partial t} = 0 \\[2mm] H_0' \overline{\psi}_2' + (V_1' + i V_2') \overline{\psi}_2' - V_3' \overline{\psi}_2' + \dfrac{\hbar}{i} \dfrac{\partial \overline{\psi}_2'}{\partial t} = 0. \end{cases} \tag{7-40'}$$

Therefore the situation is identical to that of the Newtonian equation (7-19), and the equations are invariant, independent of the coordinate system.

For these reasons, if we write the unitary matrix in (7-36) as

$$U = \begin{pmatrix} U_{11} & U_{12} \\ U_{21} & U_{22} \end{pmatrix} \tag{7-41}$$

and use these $U_{\beta\alpha}$ as $U_{\beta\alpha}$ of (7-32), then we can obtain what we want.

Furthermore since from (7-34)

$$U^\dagger = U^{-1} = \begin{pmatrix} U_{11}^* & U_{21}^* \\ U_{12}^* & U_{22}^* \end{pmatrix},$$

we can solve (7-32) and obtain

$$\overline{\psi}_\alpha = \sum_{\beta=1}^{2} \overline{\psi}'_\beta U_{\beta\alpha}^* \qquad \alpha = 1, 2. \tag{7-32'}$$

Also, remembering (7-21), we can rewrite (7-32) and (7-32') as

$$\overline{\psi}'_\beta = \sum_{\alpha=1}^{2} U_{\beta\alpha} \psi_a \qquad \beta = 1, 2 \tag{7-42}$$

$$\psi_\alpha = \sum_{\beta=1}^{2} \overline{\psi}'_\beta U_{\beta\alpha}^* \qquad \alpha = 1, 2 \tag{7-42'}$$

After this procedure we now know that a law of physics can be written in the same form in any system if we transform the components of ψ by (7-42) and (7-42') upon transformation of the coordinate system. This transformation is different from the transformation of vectors (7-9) and (7-9') not only in that there are two components, but also in that it is a unitary transformation and not an orthogonal transformation. However, there is a more fundamental difference between U and A. Our two components also covary with the rotation A in some sense, but there is a remarkable difference between the way they covary and the way vectors and tensors covary. So let us examine this point using the little time remaining.

Our U is defined by (7-33), and since σ_1', σ_2', σ_3' in (7-26') contain A_{jk}, U is also related to A_{jk}. Therefore, if we change A, U also changes. But it should be noted that even if A is given, U is not uniquely determined by (7-33). Namely, if U is a unitary matrix satisfying (7-33), then $e^{i\delta}U$, in which an arbitrary phase factor $e^{i\delta}$ is multiplied to U, is also a unitary matrix satisfying (7-33). Therefore, in order to eliminate this multivaluedness of U, we use the following method. First we obtain $|\det U|^2 = 1$ from (7-34); we can always make

$$\det U = 1 \tag{7-43}$$

by properly choosing $e^{i\delta}$. So let us impose this condition on U and try to limit its multivaluedness. Since in quantum mechanics the wave function ψ itself has the uncertainty of a phase factor, we do not have to worry about loss of generality by imposing condition (7-43). However, even if we impose condition (7-43), there still remains a two-valuedness in U. For if U satisfies (7-43), $-U$ also satisfies (7-43). For these reasons, given A_{jk}, two values $\pm U$ are allowed for U, and we can use either of them for our purpose.

Superficially, you might think that we just take one of the two U's and abandon the other. For example, $A = 1$ means no rotation of the coordinate system, and for this obviously $\sigma_j' = \sigma_j$, and therefore $U = \pm 1$. Then we just assume that it is $+1$. If we then change the value of A continuously, you might think that U varies from $+1$ continuously and is uniquely determined for arbitrary A.

In order to examine this situation, we might investigate the rotation of the coordinates around only the third axis because you probably do not want to bother with complicated mathematics.

Let us just consider the R' system, which results from rotating the R system by an angle α around the third axis. Then the rotation matrix A is

$$A(\alpha) = \begin{pmatrix} \cos\alpha & \sin\alpha & 0 \\ -\sin\alpha & \cos\alpha & 0 \\ 0 & 0 & 1 \end{pmatrix}, \tag{7-44}$$

and therefore $\sigma_1', \sigma_2', \sigma_3'$ of (7-26') are

$$\sigma_1' = \begin{pmatrix} 0 & e^{-i\alpha} \\ e^{i\alpha} & 0 \end{pmatrix} \quad \sigma_2' = \begin{pmatrix} 0 & -ie^{-i\alpha} \\ ie^{i\alpha} & 0 \end{pmatrix} \quad \sigma_3' = \begin{pmatrix} 1 & 0 \\ 0 & -1 \end{pmatrix}, \tag{7-45}$$

and after a bit of a calculation, we find that the U which satisfies (7-33) and (7-43) and becomes $+1$ for $\alpha = 0$ is given by

$$U(\alpha) = \begin{pmatrix} e^{i\alpha/2} & 0 \\ 0 & e^{-i\alpha/2} \end{pmatrix}. \tag{7-46}$$

If we look at this U, we can point out the following two things:
(I) When we apply the rotation $A(2\pi)$ then in spite of the fact that $A(2\pi) = A(0)$, $U(2\pi) \neq U(0)$, but $U(2\pi) = -U(0)$.
(II) If we specify U after the rotation $A(\alpha_1)$ as $U(\alpha_1)$ and that for $A(\alpha_2)$ as $U(\alpha_2)$, then when we apply $A(\alpha_2) \cdot A(\alpha_1)$, U is not necessarily $U(\alpha_2) \cdot U(\alpha_1)$, but in some cases when $\alpha_1 + \alpha_2$ takes a value between 2π and 4π, it is $-U(\alpha_2) \cdot U(\alpha_1)$.

From these facts we see that it is impossible to drop one of the $\pm U$ pair and make the correspondence between A and U one-to-one. Furthermore, the correspondence *product to product* does not always hold either. For these reasons we cannot consider the components ψ_1 and ψ_2 of the two-component quantity ψ as covariant quantities in the same sense as the components of vectors and tensors, and we cannot consider U as a representation of the rotation group in the previous sense. However, as is evident from (I) and (II), the discrepancy is in sign only, and it appears as though we can overcome it if we slightly extend the concept of covariant quantities and group representations.

Indeed it is possible to mathematically incorporate this U into representation theory as a "two-valued representation of the rotation group," and it was done by Hermann Weyl. Also physically, no difficulty emerges from this two-valuedness. This is because in quantum mechanics, only quantities of the form $|\psi|^2$ have physical meaning rather than ψ itself. Therefore, even if ψ itself becomes $-\psi$ after the rotation of 2π, $|\psi|^2$ goes back to its original value, and the two-valuedness of ψ never appears in physical conclusions. Therefore, ψ is fully qualified to be considered a covariant quantity.

We derived the two-valuedness for the especially simple rotation (7-44), but we can obtain the same conclusion for a general A, and this two-valuedness is the intrinsic point which discriminates $\binom{\psi_1}{\psi_2}$ from vectors and tensors. Until Pauli's theory came along, such a two-valued covariant quantity had never appeared in physics. The reason for this, I think, is that before quantum mechanics, all the covariant quantities appearing in physics were in principle themselves measurable. If so, we surely have a problem with a quantity that changes sign after a 2π rotation of the coordinate axes.

We now understand that a covariant quantity with two-valuedness plays a role in the theory, but the coordinate transformation considered so far is the usual rotation in three-dimensional space. It was found, however, that the covariant quantity thus formed is also a two-valued covariant quantity for the Lorentz transformation in Minkowski space (without having to increase the number of components). However, in this case, although det $U = 1$, U is not unitary for the transformation related to the time coordinate. For this reason, in addition to the two-component quantity which is transformed by U, we must take into account the two-component quantity which transforms by $(U^{-1})^\dagger$, and in order to discriminate these two we often write the former $\binom{\psi_1}{\psi_2}$ and the latter $\binom{\psi_1}{\psi_2}$, and if we use these, we can write Dirac's equation as simultaneous equations for $\binom{\psi_1}{\psi_2}$ and $\binom{\psi_1}{\psi_2}$.

These new quantities which are now welcomed as members of the clan of covariant quantities were christened *spinors* by Ehrenfest. We can also define second-rank spinors, third-rank spinors, and so on, and at the request of Ehrenfest, the mathematician B. L. van der Waerden developed a system of spinor algebra as opposed to tensor algebra. A remarkable thing is that spinors with even rank do not have the two-valuedness, and as a result we can use spinors with even rank to compose vectors and tensors. In this sense $\binom{\psi_1}{\psi_2}$ and $\binom{\psi_1}{\psi_2}$ are sometimes called half-vectors, and these half-vectors are the most fundamental covariant quantities from which all other covariant quantities can be composed.

For a long time—a really long time—no physicists had ever imagined that this type of covariant quantity existed. Darwin, whose name I mentioned in lecture 3, wrote a paper immediately after Dirac's in which he described his failure in an attempt to put the Dirac equation in tensor form and wrote, "It is rather disconcerting to find that apparently something has slipped through the net, so that physical quantities exist which would be [to say the least] very artificial and inconvenient to express as tensors."

Also Ehrenfest in his small paper written in 1932 wrote as follows. (The original is in German, and I translate it for you for whom German is not a strong point.) "In all measure it is truly strange that absolutely no one proposed until the work of Pauli . . . and Dirac, which is twenty years after special relativity . . . , this eerie report that a mysterious tribe by the name of spinor family inhabits isotropic [three-dimensional] space or Einstein-Minkowski [four-dimensional] world."

Today's talk was rather mathematical, and I am sorry there were so many mathematical formulas. After all, our opponent was "the mysterious tribe" which no one had caught "in the twenty years after relativity had been born." Therefore, I had no choice other than to talk in this "eerie" way. Today I am really tired.

Spin and Statistics of Elementary Particles

The Spins of Bosons Are Integers and Those of Fermions Are Half-integers

Pauli and Weisskopf's work on the possibility of spin zero particles, Dirac's hole theory, and the ensuing discovery of the positron in 1932 were deeply related in various senses; 1932 was a big year in which a spectacular avalanche of discoveries was made. For example, the neutron was also discovered in this year by J. Chadwick. And this discovery of the neutron sparked Heisenberg's theory that nuclei are composed of protons and neutrons. His proposal, namely that protons and neutrons are transformed from one to the other by exchanging an electric charge, led to Fermi's theory of β-decay, and from these two ideas Yukawa's theory of mesons was born. In 1936 C. Anderson and S. H. Neddermeyer experimentally discovered a particle that resembled Yukawa's particle.[1]

For these reasons from around 1932 to 1936, physicists' ideas about the structure of matter changed tremendously. Namely, the world of elementary particles turned out to be much more diverse than had been considered before 1932. Until then the only elementary particles which had been considered were the proton, the electron, and the photon, but to these the positron, neutron, and meson were now added. Furthermore, although experimentally not yet proven to exist, particles like the neutrino and the graviton seemed theoretically possible. Therefore, it was unimaginable what other elementary particles might be found. This change of thought on elementary particles came about around 1936.

If the world of elementary particles is indeed so diverse, something like a central pillar or axis is required which connects all the complicated relations and straightens out the logical structure. Pauli's theory on the relation between the spin and statistics of elementary particles emerged from this background. In this theory, published in 1940, Pauli derived, as our subtitle says, that the only particles with integral values of spin are bosons, and the only ones with

---

1. Actually this was a μ meson which is a decay product of Yukawa's π meson.

half-integral spin are fermions. The starting point of this idea was none other than the Pauli-Weisskopf theory, and therefore their paper was not just of passing significance as a retaliation against Dirac but also very actively directed the development of physics thereafter.

Today I would like to talk about this work of Pauli's. To begin, let us take the Klein-Gordon particle as a representative of particles with integral spin and the Dirac particle as a representative of particles with half-integral spin. If you remember lecture 6, when we quantize the field, a particle appropriate to the field appears, and the field can be quantized in two ways. One of them is

$$\Psi(\mathbf{x})\Pi(\mathbf{x'}) - \Pi(\mathbf{x'})\Psi(\mathbf{x}) = i\hbar\,\delta(\mathbf{x} - \mathbf{x'}) \qquad (8\text{-}1)_-$$

and the other

$$\Psi(\mathbf{x})\Pi(\mathbf{x'}) + \Pi(\mathbf{x'})\Psi(\mathbf{x}) = i\hbar\,\delta(\mathbf{x} - \mathbf{x'}). \qquad (8\text{-}1)_+$$

You must remember that a boson appears when $(8\text{-}1)_-$ is used and a fermion appears when $(8\text{-}1)_+$ is used. Therefore, Pauli showed first that it is impossible to quantize the Klein-Gordon field with $(8\text{-}1)_+$ and the Dirac field with $(8\text{-}1)_-$.

Now let us start from the Klein-Gordon field, for which the field equation is

$$\left[\left(-\frac{\hbar}{ic}\frac{\partial}{\partial t}\right)^2 - \sum_{r=1}^{3}\left(\frac{\hbar}{i}\frac{\partial}{\partial x_r}\right)^2 - m^2 c^2\right]\Psi(\mathbf{x}, t) = 0. \qquad (8\text{-}2)$$

[We do not treat the case where more than one field coexist today. Therefore the electromagnetic potential A_0, $\mathbf{A}$ does not appear in (8-2).] Around 1926–27, before the Dirac equation hit the scene, when people were not yet certain whether Ψ is a wave in configuration space or in three-dimensional space, many already thought that (8-2) was the relativistic field equation for an electron wave (and therefore leaned toward the three-dimensional idea) and discussed its structure. These people (who included not only Klein and Gordon but of course Schrödinger, who favored using the three-dimensional picture) engaged in the work of applying to this equation the general method for field theory which had been established in the theory of relativity. That is to say, they started from the Lagrangian of the field, derived the field equation, and determined the energy-momentum tensor for the matter field. If the electromagnetic field is incorporated into this procedure, which we are not doing today, the mutual interaction between the matter field and the electromagnetic field is determined, and the electric current vector associated with the matter field is obtained. Furthermore, it is possible to derive the canonical variables for the field (in the sense of Hamilton) from the Lagrangian although this was not done at that time because it was not very necessary.

According to these studies, the energy density (which is the time-time component of the energy-momentum tensor) of the Klein-Gordon field is

$$H = c^2\hbar^2 \left(\frac{1}{c^2} \left| \frac{\partial \Psi}{\partial t} \right|^2 + \sum_{r=1}^{3} \left| \frac{\partial \Psi}{\partial x_r} \right|^2 + \kappa^2 |\Psi|^2 \right), \qquad (8\text{-}3)$$

and the charge density is the time component of the electric current vector, given by

$$\rho = \frac{e\hbar}{2mc^2} \frac{1}{i} \left(\Psi^* \frac{\partial \Psi}{\partial t} - \frac{\partial \Psi^*}{\partial t} \Psi \right). \qquad (8\text{-}4)$$

Here κ in (8-3) is

$$\kappa = \frac{mc}{\hbar}. \qquad (8\text{-}3')$$

It was further found that if we consider Ψ and Ψ^* as canonical coordinates, the momenta Π and Π^* conjugate to them are

$$\Pi = \hbar^2 \frac{\partial \Psi^*}{\partial t} \qquad \Pi^* = \hbar^2 \frac{\partial \Psi}{\partial t}. \qquad (8\text{-}5)$$

(Let me remind you that for the time being the field is not quantized.)

Now let us examine what the results obtained mean. The first conclusion from (8-3) is that the right-hand side of H is a sum of terms of the form $|\ldots|^2$ and therefore is always positive, namely, energy is always positive. On the other hand, we can conclude that the charge density may be positive or negative. For if Ψ is one of the solutions of (8-2), then obviously Ψ^* also satisfies (8-2), and therefore the expression (8-4) in which Ψ is replaced by Ψ^* [for Ψ^* we use $(\Psi^*)^* = \Psi$] is also a possible charge density; if ρ is a possible density, then $-\rho$ also is a possible density. Thus we obtain the conclusion that ρ could be either positive or negative. [I told you two lectures back that Dirac disliked this point that the electric charge may be positive or negative, and he used exactly this argument to expose this "fault" of (8-2).]

What happens for the Dirac equation? As given in lecture 3, the Dirac equation is

$$\left[\left(-\frac{\hbar}{ic} \frac{\partial}{\partial t} \right) - \sum_{r=1}^{3} \alpha_r \left(\frac{\hbar}{i} \frac{\partial}{\partial x_r} \right) - \alpha_0 mc \right] \Psi(\mathbf{x}, t) = 0. \qquad (8\text{-}6)$$

For the four matrices α_0, α_r Dirac used

$$\alpha_1 = \begin{pmatrix} 0 & 0 & 0 & 1 \\ 0 & 0 & 1 & 0 \\ 0 & 1 & 0 & 0 \\ 1 & 0 & 0 & 0 \end{pmatrix} \quad \alpha_2 = \begin{pmatrix} 0 & 0 & 0 & -i \\ 0 & 0 & i & 0 \\ 0 & -i & 0 & 0 \\ i & 0 & 0 & 0 \end{pmatrix}$$

$$\alpha_3 = \begin{pmatrix} 0 & 0 & 1 & 0 \\ 0 & 0 & 0 & -1 \\ 1 & 0 & 0 & 0 \\ 0 & -1 & 0 & 0 \end{pmatrix} \quad \alpha_0 = \begin{pmatrix} 1 & 0 & 0 & 0 \\ 0 & 1 & 0 & 0 \\ 0 & 0 & -1 & 0 \\ 0 & 0 & 0 & -1 \end{pmatrix} \tag{8-7}$$

as I told you earlier.

For the Dirac equation also we can apply the standard method of relativity using the Lagrangian. According to this, the energy density of the field is

$$H = \frac{1}{2} \frac{\hbar}{i} \left(\Psi^* \frac{\partial \Psi}{\partial t} - \frac{\partial \Psi^*}{\partial t} \Psi \right), \tag{8-8}$$

and the charge density is

$$\rho = -e\Psi^*\Psi, \tag{8-9}$$

where the charge of the electron is taken as $-e$. Furthermore, the canonical momentum Π conjugate to Ψ is given by

$$\Pi = i\hbar\Psi^*. \tag{8-10}$$

[Unlike in (8-5), we cannot consider Ψ^* and Ψ as independent coordinates, for Ψ^* is already a momentum conjugate to Ψ as shown in (8-10).]

Now what happens to the sign of H and ρ? Let us first change i into $-i$ in (8-6). Then that equation becomes

$$\left[\left(\frac{\hbar}{ic} \frac{\partial}{\partial t} \right) - \sum_{r=1}^{3} \alpha_r^* \left(-\frac{\hbar}{i} \frac{\partial}{\partial x_r} \right) - \alpha_0^* mc \right] \Psi^*(\mathbf{x}, t) = 0.$$

Here α^* is a matrix in which i in the matrix elements of α is changed to $-i$. If we now look at (8-7), it is only α_2 whose elements contain i. Therefore, we can rewrite this formula as

$$-\left[\left(-\frac{\hbar}{ic} \frac{\partial}{\partial t} \right) - \sum_{r=1,3} \alpha_r \left(\frac{\hbar}{i} \frac{\partial}{\partial x_r} \right) + \alpha_2 \left(\frac{\hbar}{i} \frac{\partial}{\partial x_2} \right) + \alpha_0 mc \right] \Psi^*(\mathbf{x}, t) = 0.$$

$$\tag{8-6}^*$$

The expression inside the brackets of (8-6)* is different from that inside the brackets of (8-6). Therefore, unlike the case of the Klein-Gordon equation, Ψ^* is not a solution of the Dirac equation. However, if we introduce the matrix

$$
C = \begin{pmatrix} 0 & 0 & 0 & -1 \\ 0 & 0 & 1 & 0 \\ 0 & 1 & 0 & 0 \\ -1 & 0 & 0 & 0 \end{pmatrix},
\tag{8-11}
$$

then we can derive from (8-7) that

$$
C\alpha_1 = \alpha_1 C, \quad C\alpha_3 = \alpha_3 C, \quad C\alpha_2 = -\alpha_2 C, \quad C\alpha_0 = -\alpha_0 C. \tag{8-11'}
$$

Therefore, if we multiply this C to the left-hand side of the brackets of (8-6)* and move C to the right-hand side of the brackets using (8-11'), then we obtain

$$
\left[\left(-\frac{\hbar}{ic} \frac{\partial}{\partial t} \right) - \sum_{r=1}^{3} \alpha_r \left(\frac{\hbar}{i} \frac{\partial}{\partial x_r} \right) - \alpha_0 mc \right] C\Psi^*(\mathbf{x}, t) = 0. \tag{8-6}_C^*
$$

From this we obtain the conclusion that "if Ψ is a solution of (8-6), then $C\Psi^*$ is also a solution of (8-6)."

We can now use this conclusion for examining the signs of H and ρ. First, if we write (8-8) using the components Ψ_α ($\alpha = 1, 2, 3, 0$), note

$$
H = \frac{1}{2} \frac{\hbar}{i} \sum_\alpha \left(\Psi_\alpha^* \frac{\partial \Psi_\alpha}{\partial t} - \frac{\partial \Psi_\alpha^*}{\partial t} \Psi_\alpha \right),
$$

and use $\sum_\beta C_{\alpha\beta} \Psi_\beta^*$ for Ψ_α and $\sum_\beta C_{\alpha\beta}^* \Psi_\beta$ for Ψ_α^* in this expression, then since $\sum_\beta C_{\alpha\beta}^* C_{\alpha'\beta} = \delta_{\alpha\alpha'}$, we immediately obtain

$$
H_{\Psi \to C\Psi^*} = -H. \tag{8-12}
$$

Therefore, we obtain the conclusion that if H is a possible energy density, then $-H$ is also a possible energy density, and therefore energy could be either positive or negative. As for ρ, since $\Psi^* \Psi = \sum_\alpha \Psi_\alpha^* \Psi_\alpha$ is always positive for any Ψ, ρ always has the same sign as $-e$. It was this latter property which convinced Dirac that this equation is suitable for describing an electron because he was working before the discovery of the positron. However, he faced the dilemma that H could be either positive or negative. By the way, when the field is complex in character, the transformation which makes from a solution Ψ of a field equation another solution which is its complex conjugate (i.e., Ψ^* for the Klein-Gordon equation and $C\Psi^*$ for the Dirac equation) is called "charge conjugation." This transformation would later play a great role in elementary particle physics.

In summary, so far we have obtained "complementary" conclusions that

$$\left\{ \begin{array}{l} \text{for a Klein-Gordon field} \\[4pt] \quad \text{(i) energy is always positive} \\[4pt] \quad \text{(ii) electric charge may be positive or negative;} \\[4pt] \text{for a Dirac field} \\[4pt] \quad \text{(i) electric charge is always a positive multiple of } -e \\[4pt] \quad \text{(ii) energy can be positive or negative.} \end{array} \right. \tag{8-13}$$

So far we have not quantized the field. First we must consider exactly how we want to do it. As I told you earlier, there are two methods of quantization, one for bosons and the other for fermions. What we must consider is which quantization to carry out for the Klein-Gordon field and which for the Dirac field. As I told you two lectures ago, Dirac resorted to Pauli's exclusion principle to avoid the difficulty of negative energy. Therefore, we obtain only nonsensical results if we use the quantization of bosons for the Dirac field. On the other hand, Pauli and Weisskopf pointed out that quantizing the Klein-Gordon field using the bosonic commutation relation does not introduce any inconsistency but using the fermionic commutation relation does. Therefore, in summary,

$$\left\{ \begin{array}{l} \text{for a Klein-Gordon field} \\[4pt] \quad \text{(i) bosonic commutation relation is OK} \\[4pt] \quad \text{(ii) fermionic commutation relation is no good;} \\[4pt] \text{for a Dirac field} \\[4pt] \quad \text{(i) fermionic commutation relation is OK} \\[4pt] \quad \text{(ii) bosonic commutation relation is no good,} \end{array} \right. \tag{8-14}$$

and these relations are also "complementary."

We omit the proof of (8-14) here because we shall talk later about the work in which Pauli showed that in general

$\left\{\begin{array}{l} \text{for a tensor field} \\ \quad \text{(i) bosonic commutation relation is OK} \\ \quad \text{(ii) fermionic commutation relation is no good;} \\ \text{for a spinor field} \\ \quad \text{(i) fermionic commutation relation is OK} \\ \quad \text{(ii) bosonic commutation relation is no good.} \end{array}\right.$ (8-14')

Also, as for the generalization of (8-13), i.e.,

$\left\{\begin{array}{l} \text{for a tensor field} \\ \quad \text{(i) energy is always positive} \\ \quad \text{(ii) electric charge may be positive or negative;} \\ \text{for a spinor field} \\ \quad \text{(i) electric charge is always a positive multiple of } -e \\ \quad \text{(ii) energy can be positive or negative,} \end{array}\right.$ (8-13')

Pauli proved that (ii) are always valid. His assistant M. Fierz discussed statements (i), and according to him, they are not generally true but are valid for spins of 0, 1/2, and 1. For now, we shall postpone our consideration of quantization and start to discuss how these conclusions were derived.

Pauli derived relations (8-13') from very general considerations, actually from only three assumptions:
  (a) the quantity representing the field is either a tensor or a spinor;
  (b) the field equation is a covariant, linear, homogeneous, differential equation in (x, y, z, ct);
  (c) the general solution of this equation can be expressed by a superposition of plane waves $e^{i(\mathbf{k}\cdot\mathbf{x}-ck_0t)}$, where $k_0 = \pm\sqrt{\mathbf{k}^2 + \kappa^2}$.
As seen below, his conclusion is more general than (8-13').

Now let me introduce to you Pauli's idea for when the field quantity is a tensor. We first classify the tensors U into those of even rank and those of odd rank, and write the general even-rank tensor as U^e and the general odd-rank tensor as U^o. If we discuss individual even-rank tensors rather than general ones, we shall write $U_1^e, U_2^e, \ldots$ with a superscript "e". Likewise, individual odd-rank tensors will be $U_1^o, U_2^o, \ldots$ These U can be either real or complex, and in the latter case usually U and U^* together are used to describe the field, but we do not have to assume that. Next, we write a product of two tensors as $U \times U$. (A direct product to which a constant is multiplied or which is contracted, symmetrized, or antisymmetrized an arbitrary number of times is summarily called a *product*.) Then the product of two even-rank tensors is always of even rank, and the product of two odd-rank tensors is always of even rank, and the product of an odd- and an even-rank tensor is always of odd rank. So we write

$$U^e \times U^e = U^e, \qquad U^o \times U^o = U^e, \qquad U^e \times U^o = U^o. \quad (8\text{-}15)$$

Next remember that the differential operator

$$\nabla^o \equiv \frac{1}{i} \left(\frac{\partial}{\partial x}, \frac{\partial}{\partial y}, \frac{\partial}{\partial z}, \frac{1}{c}\frac{\partial}{\partial t} \right) \qquad (8\text{-}16)$$

can be considered to be a vector. I put a superscript ^{o} on ∇ because the vector is an odd-rank tensor. We can then make all sorts of tensorlike differential operators by $\nabla^o \times \nabla^o$, $\nabla^o \times \nabla^o \times \nabla^o$, $\ldots$, and we denote these tensors in general as D. We then have from (8-15), $\nabla^o = D^o$, $\nabla^o \times \nabla^o = D^e$, $\nabla^o \times \nabla^o \times \nabla^o = D^o$, $\ldots$ Also, for D^e and D^o operating on an arbitrary tensor we have

$$D^e U^e = U^e, \quad D^o U^o = U^e, \quad D^e U^o = U^o, \quad D^o U^e = U^o. \quad (8\text{-}16')$$

So far we have not considered the field equation, and therefore (8-15) and (8-16') are valid for any U. What will happen if we consider the field equation? There is no further assumption necessary other than that the equation satisfy (b) and (c). For example, we do not have to assume that the field must be described by a single tensor, nor must we assume that the equation is first-order. (Pauli may have had Dirac in mind when he deliberately says that it does not have to be first-order.) So we assume that the field is described by M quantities U^e — $U_1^e, U_2^e, \ldots, U_M^e$ — and N quantities U^o — $U_1^o, U_2^o, \ldots, U_N^o$. Then the field equations must be a system of $M + N$ simultaneous differential equations, but since they have to satisfy the covariance requirement (b), each of the equations based on (8-16') must be of the form

$$\sum_{m=1}^{M} D_m^e U_m^e = \sum_{n=1}^{N} D_n^o U_n^o, \quad \sum_{m=1}^{M} D_m^o U_m^e = \sum_{n=1}^{N} D_n^e U_n^o. \quad (8\text{-}17)$$

Furthermore, since because of assumption (c) the solutions are of the form

$$U_m^e = \overline{U}_m^e(K)e^{i(\mathbf{k}\cdot\mathbf{x}-ck_0 t)}, \qquad U_n^o = \overline{U}_n^o(K)e^{i(\mathbf{k}\cdot\mathbf{x}-ck_0 t)}, \quad (8\text{-}18)$$

by substituting these into (8-17),

$$\sum_{m=1}^{M} K_m^e \overline{U}_m^e(K) = \sum_{n=1}^{N} K_n^o \overline{U}_n^o(K),$$

$$\sum_{m=1}^{M} K_m^o \overline{U}_m^e(K) = \sum_{n=1}^{N} D_n^e \overline{U}_n^o(K)$$

(8-17')

are derived. I wrote in (8-18) $\overline{U}_m^e(K)$ or $\overline{U}_n^o(K)$ in order explicitly to show that they are the amplitudes of the plane waves having the propagation vector

$$K \equiv (k_x, k_y, k_z, k_0). \tag{8-19}$$

In (8-17'), K_m^e and K_n^e are both even-rank tensors formed from vector (8-19) by $K \times K$, $K \times K \times K \times K, \ldots$, and K_m^o and K_n^o are odd-rank tensors formed from K, $K \times K \times K, \ldots$ In general, K^e is an even function of k_x, k_y, k_z, k_0 (i.e., a function which does not change value by $k_x \to -k_x, k_y \to -k_y, k_z \to -k_z, k_0 \to -k_0$), and K^o is an odd function (a function whose sign changes).

For the collection or class of amplitudes $\overline{U}_m^e(K)(m = 1, 2, \ldots, M)$ and $\overline{U}_n^o(K)(n = 1, 2, \ldots, N)$ we have $M + N$ simultaneous algebraic equations of the form (8-17'). We can obtain the possible class of amplitudes by solving the equations, but in general there will be more than one solution, and therefore there must be more than one class of plane waves that satisfy the field equation. (In this case, for a certain class $k_0 = +\sqrt{\mathbf{k}^2 + k^2} > 0$, and for other classes $k_0 = -\sqrt{\mathbf{k}^2 + k^2} < 0$.) From the facts pointed out earlier that K^e is an even function of $\mathbf{k}$ and k_0 and that K^o is an odd function, we obtain the following important conclusion.

Let us assume that we have obtained a class of solutions $\overline{U}_m^e(K)(m = 1, 2, \ldots, M)$ and $\overline{U}_n^o(K)(n = 1, 2, \ldots, N)$, where K is the value of the propagation vector. Then the class of $\overline{U}_m^e(-K)$ and $\overline{U}_n^o(-K)$, given by

$$\overline{U}_m^e(-K) = \overline{U}_m^e(K), \qquad \overline{U}_n^o(-K) = -\overline{U}_n^o(K), \tag{8-20}$$

is also a class of solutions for the propagation vector $-K$. This is easily seen if we replace K in (8-17') with $-K$ and use (8-20). If the original plane wave (8-18) belonged to the class with $k_0 > 0$ (or $k_0 < 0$), then the plane waves of (8-20), namely

$$\overline{U}_m^e(K)e^{i[(-\mathbf{k})\cdot\mathbf{x} - c(-k_0)t]}, \qquad -\overline{U}_n^o(K)e^{i[(-\mathbf{k})\cdot\mathbf{x} - c(-k_0)t]} \tag{8-18'}$$

belong to the class of $k_0 < 0$ (or $k_0 > 0$). In this sense (8-20) means the transformation from classes with $k_0 > 0$ to classes with $k_0 < 0$ or from classes with $k_0 < 0$ to classes with $k_0 > 0$. Here since (8-20) is symmetric with respect to K and $-K$, there is a one-to-one correspondence between the *members of classes* in this transformation (therefore the classes with $k_0 > 0$ and the classes with $k_0 < 0$ are equal in number). Pauli expressed this transformation by the notation

$$K \rightarrow -K, \qquad U^e \rightarrow U^e, \qquad U^o \rightarrow -U^o. \tag{8-21}$$

So far, we have discussed plane-wave solutions, but from assumption (c) we can produce a general solution by their superposition. First, we take a linear combination of plane waves (8-18) using arbitrary coefficients $\alpha(\mathbf{k})$ for each class and make a wave packet

$$\Psi_m^e(\mathbf{x}, t) = \sum_{\mathbf{k}} \overline{U}_m^e(K)\alpha(\mathbf{k})e^{i(\mathbf{k}\cdot\mathbf{x}-ck_0t)}, \quad m = 1, 2, \ldots, M$$

$$\Psi_n^o(\mathbf{x}, t) = \sum_{\mathbf{k}} \overline{U}_n^o(K)\alpha(\mathbf{k})e^{i(\mathbf{k}\cdot\mathbf{x}-ck_0t)}, \quad n = 1, 2, \ldots, N. \tag{8-22}$$

[Here $\alpha(\mathbf{k})$ may be different for different classes.] We next sum the wave packets formed for each class over all the classes. Then the wave packet given by the sum

$$U_m^e(\mathbf{x}, t) = \sum_{\text{class}} \Psi_m^e(\mathbf{x}, t), \quad m = 1, 2, \ldots, M$$

$$U_n^o(\mathbf{x}, t) = \sum_{\text{class}} \Psi_n^o(\mathbf{x}, t), \quad n = 1, 2, \ldots, N, \tag{8-23}$$

is the general solution.

In preparation for later discussion, I mention the following. First we form (8-18') from (8-18) by a Pauli transformation. Then since (8-18') satisfies the field equation, the wave packet

$$\Psi_m'^e(\mathbf{x}, t) = \sum_{\mathbf{k}} \overline{U}_m^e(K)\alpha(\mathbf{k})e^{i[(-\mathbf{k})\cdot\mathbf{x}-c(-k_0)t]}$$

$$\Psi_n'^o(\mathbf{x}, t) = -\sum_{\mathbf{k}} \overline{U}_n^o(K)\alpha(\mathbf{k})e^{i[(-\mathbf{k})\cdot\mathbf{x}-c(-k_0)t]} \tag{8-22'}$$

also satisfies the field equation, and therefore the wave packet formed from them,

$$U'^{e}_{m}(\mathbf{x}, t) = \sum_{\text{class}} \Psi'^{e}_{m}(\mathbf{x}, t)$$

$$U'^{o}_{n}(\mathbf{x}, t) = \sum_{\text{class}} \Psi'^{o}_{n}(\mathbf{x}, t) \tag{8-23'}$$

satisfies the field equation. We also call this transformation which produces U' from U a Pauli transformation.

Now in relativistic field theory, covariant quadratic forms or bilinear forms composed from the quantities of the field and their derivatives play important roles. The energy-momentum tensor is a good example. Also if the matter field interacts with an electromagnetic field, then the electric current vector is also an example. As I told you at the beginning, the energy-momentum tensor is a second-rank tensor, and its time-time component gives the energy density, and the electric current vector is a first-rank tensor, and its time component gives the electric charge density. Let us examine how much can be concluded from the very general discussion just given on the properties of the first-rank and second-rank tensors composed from the quantities U^{e}_{m} and U^{o}_{n}, which describe the field, and their derivatives.

Let us start from the first-rank tensor. A first-rank tensor is an odd-rank tensor, and therefore according to (8-15) it must be of the form $U^{e} \times U^{o}$. We then use (8-16'), but let us notice first that since the argument from now on is only for the distinction of whether the tensor is even or odd, D^{e} in (8-16') does not play any role. That is to say we can see from (8-16') that D^{e} operating on U^{e} or U^{o} does not change the parity of U^{e} or U^{o} at all. Therefore, if we are considering only the parity of U, then all D^{e} can be summarily represented by 1. Using the same argument, if the problem is only the parity of U, all D^{o} can be summarily represented by ∇^{o}. In this sense, let us write (8-16') as

$$1U^{e} = U^{e}, \qquad \nabla^{o}U^{o} = U^{e} \tag{8-16''}$$
$$1U^{o} = U^{o}, \qquad \nabla^{o}U^{e} = U^{o}.$$

If we do this, the two equations on the left are trivial, and it is not necessary to write them down.

After this remark, we come back to the story of first-rank tensors (or in general, odd-rank tensors). As mentioned earlier, they are of the form $U^{e} \times U^{o}$, and according to (8-16''), they are one of $U^{e} \times \nabla^{o}U^{e}$, $\nabla^{o}U^{o} \times U^{o}$, or $U^{e} \times U^{o}$. Since the order of multiplication does not affect the parity, we omit the "$\times$" and simply write them as $U^{e}\nabla^{o}U^{e}$, $U^{o}\nabla^{o}U^{o}$, $U^{e}U^{o}$. Then the most general form of the first-rank tensor that can be generated from U^{e} and U^{o} or their derivatives is

$$S = U^e \nabla^o U^e + U^o \nabla^o U^o + U^e U^o. \qquad (8\text{-}24)$$

We then use (8-23) for U^e and U^o on the right-hand side. We obtain

$$S = \sum_{\text{class}} \sum_{\text{class}'} \left(\sum_{m,m'} \Psi_m^e \nabla^o \Psi_{m'}^e + \sum_{n,n'} \Psi_n^o \nabla^o \Psi_{n'}^o + \sum_{m,n'} \Psi_m^e \Psi_{n'}^o \right). \qquad (8\text{-}25)$$

Here the meaning of $\sum_{\text{class}}, \sum_{\text{class}'}$ is self-explanatory, but I might just remind you that on the right-hand side Ψ_m^e and Ψ_n^o belong to the same class, and so do $\Psi_{m'}^e$ and $\Psi_{n'}^o$, but Ψ_m^e and $\Psi_{m'}^e$ (likewise Ψ_n^o and $\Psi_{n'}^o$) may or may not belong to the same class.

Now let us finally discuss the sign of S. In order to do this, we calculate S using (8-22) and (8-22'), and compare the results. First if we denote

$$\overline{U}_m^e(K)\alpha(\mathbf{k}) \equiv A_m^e(K), \qquad \overline{U}_n^o(K)\alpha(\mathbf{k}) \equiv A_n^o(K) \qquad (8\text{-}26)$$

and calculate S using (8-22), we obtain

$$S(\mathbf{x}, t) = \sum \sum{}' \left[A_m^e(K)K'A_{m'}^e(K') + A_n^o(K)K'A_{n'}^o(K') + A_m^e(K)A_{n'}^o(K') \right]$$
$$\times e^{i[(\mathbf{k}+\mathbf{k}')\cdot\mathbf{x}-c(k_0+k_0')t]}, \qquad (8\text{-}27)$$

but when (8-22') is used, S' is

$$S(\mathbf{x}, t) = -\sum \sum{}' \left[A_m^e(K)K'A_{m'}^e(K') + A_n^o(K)K'A_{n'}^o(K') + A_m^e(K)A_{n'}^o(K') \right]$$
$$\times e^{i[(-\mathbf{k}-\mathbf{k}')\cdot\mathbf{x}-c(-k_0-k_0')t]}. \qquad (8\text{-}27')$$

Here $\sum$ is a combination of $\sum_{\text{class}}, \sum_m$ (or $\sum_n$), and $\sum_{\mathbf{k}}$, and $\sum'$ is a combination of $\sum_{\text{class}'}, \sum_{m'}$ (or $\sum_{n'}$), and $\sum_{\mathbf{k}'}$. If we note

$$e^{i[(-\mathbf{k}-\mathbf{k}')\cdot\mathbf{x}-c(-k_0-k_0')t]} = e^{i[(\mathbf{k}+\mathbf{k}')\cdot(-\mathbf{x})-c(k_0+k_0')(-t)]},$$

we derive the following conclusion from (8-27) and (8-27'). It is, "If the value of S at $\mathbf{x}$, t using the wave packet (8-23) is positive (or negative), the value of S at $-\mathbf{x}$, $-t$ using the wave packet (8-23') is negative (or positive)." This leads to the compelling conclusion that in the case of a tensor field, S cannot take only positive values or only negative values.

This conclusion was derived for the first-rank tensor, and it naturally applies to the electric current vector as a special case. Thus we can conclude that the charge density in a tensor field may have either sign.

What happens to the energy-momentum tensor? This is a case of a second-rank tensor, whose most general form is

$$T = U^e U^e + U^o U^o + U^e \nabla^o U^o. \tag{8-28}$$

Following an argument analogous to that for S,

$$T(\mathbf{x}, t) = \sum \sum' \left[A_m^e(K) A_{m'}^e(K') + A_n^o(K) A_{n'}^o(K') + A_m^e(K) K' A_{n'}^o(K') \right]$$
$$\times e^{i[(\mathbf{k}+\mathbf{k}')\cdot\mathbf{x}-c(k_0+k_0')t]}, \tag{8-29}$$

and

$$T'(\mathbf{x}, t) = \sum \sum' \left[A_m^e(K) A_{m'}^e(K') + A_n^o(K) A_{n'}^o(K') + A_m^e(K) K' A_{n'}^o(K') \right]$$
$$\times e^{i[(\mathbf{k}+\mathbf{k}')\cdot(-\mathbf{x})-c(k_0+k_0')(-t)]}, \tag{8-29'}$$

are obtained, leading to the conclusion that "the value of T at $(\mathbf{x}, t)$ using the wave packet (8-23), and the value of T at $(-\mathbf{x}, -t)$ using the wave packet (8-23') are equal to one another." We cannot conclude, however, that T is always positive, but at least such a possibility is not excluded.

Next, what would happen in the case of a spinor field? Unfortunately, we do not have enough time to examine this situation thoroughly, nor have we covered in sufficient detail the spinor algebra needed for this argument. Therefore please bear with me for just giving the conclusion. We find that in the spinor field, T is not of definite sign. As for S, we cannot conclude that S is always positive, but at least the possibility is not excluded. From this we find that in the spinor field the energy density can be either positive or negative and its space integral, namely energy, can also be positive or negative.

This was Pauli's argument in deriving (ii) of (8-13'). Earlier, we derived (8-13) for the Klein-Gordon field and the Dirac field using the charge conjugation transformation. The characteristic of Pauli's general argument is that it does not use this transformation but instead uses the Pauli transformation (8-21). As a result of this, Pauli's conclusion about the electric current vector is applicable in general to all odd-rank tensors, and his conclusion about the energy-momentum tensor applies to all even-rank tensors and to scalar fields. It cannot be deduced that T for the tensor field and S for the spinor field are always positive from Pauli's argument alone. Indeed as I mentioned earlier, Fierz showed that it is valid only when the spin is 0, 1/2, or 1. Nevertheless, it is very instructive to realize that such a profound conclusion can be derived only from the assumption that "physical quantities must be covariant quantities," and in the process tensor and spinor algebra demonstrate their power.

We now move on to field quantization. You will see that the concept of covariance plays a role here, also. However in this discussion, the usual form of the commutator (or anticommutator) presents a knotty problem. For example, if you look at (8-1)_, we have on the left-hand side $\Psi(\mathbf{x})$ and $\Pi(\mathbf{x}')$, but these mean Ψ and Π at different positions but at the same instant in time, so strictly they should be written $\Psi(\mathbf{x}, t)$, $\Pi(\mathbf{x}', t)$. Such a different treatment of the space variable and the time variable is not relativistic, and therefore we cannot draw any conclusion using the covariance of four-dimensional space. Therefore, we must find for our discussions a commutation relation which relates the field at different positions and times, namely, in the form of

$$\Psi(\mathbf{x}, t)\Pi(\mathbf{x}', t') - \Pi(\mathbf{x}', t')\Psi(\mathbf{x}, t) = F(\mathbf{x} - \mathbf{x}', t - t').$$

And the left- and right-hand sides of this equation should be covariant.

Does such an F indeed exist? And if it does, what are its properties? The simplest Klein-Gordon field must give a hint. Therefore let us try that.

First let us point out that the canonical commutation relations for the Klein-Gordon field are obtained by substituting (8-5) into (6-14'):

$$\Psi(\mathbf{x}, t)\frac{\partial \Psi^{\dagger}(\mathbf{x}', t)}{\partial t} - \frac{\partial \Psi^{\dagger}(\mathbf{x}', t)}{\partial t}\Psi(\mathbf{x}, t) = \frac{i}{\hbar}\delta(\mathbf{x} - \mathbf{x}')$$

$$\Psi^{\dagger}(\mathbf{x}, t)\frac{\partial \Psi(\mathbf{x}', t)}{\partial t} - \frac{\partial \Psi(\mathbf{x}', t)}{\partial t}\Psi^{\dagger}(\mathbf{x}, t) = \frac{i}{\hbar}\delta(\mathbf{x} - \mathbf{x}') \qquad (8\text{-}30)$$

(Since Ψ is a q-number, we use $\Psi^{\dagger}$ instead of Ψ^*.) In addition to these, there are relations in which Ψ and $\Psi^{\dagger}$, $\partial\Psi/\partial t$ and $\partial\Psi^{\dagger}/\partial t$, $\partial\Psi/\partial t$ and Ψ, and $\partial\Psi^{\dagger}/\partial t$ and $\Psi^{\dagger}$ all commute. As I said earlier, one unsatisfactory point of these commutation relations is that t in Ψ and t in $\Psi^{\dagger}$ are the same values, but note that since (8-30) and its additional relations are known, we can calculate from them the relations for commutators of the type

$$\Psi(\mathbf{x}, t)\Psi^{\dagger}(\mathbf{x}', t') - \Psi^{\dagger}(\mathbf{x}', t')\Psi(\mathbf{x}, t)$$

using the fact that Ψ and $\Psi^{\dagger}$ change with time satisfying (8-2). Namely, if for example we put

$$\Psi(\mathbf{x}, t)\Psi^{\dagger}(\mathbf{x}', t') - \Psi^{\dagger}(\mathbf{x}', t')\Psi(\mathbf{x}, t) \equiv \frac{i}{\hbar}\Delta(\mathbf{x} - \mathbf{x}', t - t'), \qquad (8\text{-}31)$$

then we see

(I) since $\Psi(\mathbf{x}, t)$ on the left-hand side satisfies the Klein-Gordon equation, $\Delta(\mathbf{x} - \mathbf{x}', t - t')$ on the right-hand side must also satisfy the equation;

(II) since Ψ and $\Psi^\dagger$ commute at $t = t'$, $\Delta(\mathbf{x} - \mathbf{x}', t - t')$ on the right-hand side is zero at $t' = t$;

(III) because of (8-30), for $t' = t$, we see $\partial \Delta(\mathbf{x} - \mathbf{x}', t - t')/\partial t|_{t'=t} = \delta(\mathbf{x} - \mathbf{x}')$.

Therefore, the function $\Delta(\mathbf{x} - \mathbf{x}', t - t')$ may be obtained by solving the Klein-Gordon equation using (II) and (III) as initial conditions. Then Δ is uniquely determined.

For our purpose it is not necessary to write down the functional form of Δ explicitly. Here we just point out the following three properties of Δ obtained in this way:

(A) Δ is a real-valued function of $\mathbf{x}^2 - c^2 t^2$;

(B) $\Delta(\mathbf{x} - \mathbf{x}', t - t')$ is symmetric with respect to interchange of the space variables $\mathbf{x}$ and $\mathbf{x}'$ and antisymmetric with respect to interchange of the time variables t and t'; and

(C) $\Delta(\mathbf{x} - \mathbf{x}', t - t')$ is zero in the region $-|\mathbf{x} - \mathbf{x}'| < c(t - t') < |\mathbf{x} - \mathbf{x}'|$.

Furthermore, since the left-hand side of (8-31) is a scalar, Δ is also a scalar, but from (A) we can conclude that Δ is not just a scalar but an invariant scalar function. Here by an invariant scalar function I mean a function $f(x, y, z, ct)$ which satisfies not simply

$$f(x, y, z, ct) = \overline{f}(x', y', z', ct')$$

when the Lorentz transformation

$$x, y, z, ct \rightarrow x', y', z', ct'$$

is made, but one for which the functional forms of f and $\overline{f}$ are identical. [The meaning of the notation $\overline{f}$ was given in the previous lecture. See (7-20').] We now see that the right-hand side of the commutation relation (8-31) has the identical form in any systems that are related by Lorentz transformations. This last remark is essential. If Δ has different forms in different Lorentz systems, then the requirement that all Lorentz systems be equivalent will not be satisfied.

In addition to the commutation relation (8-31), we can derive

$$\Psi(\mathbf{x}, t)\Psi(\mathbf{x}', t') - \Psi(\mathbf{x}', t')\Psi(\mathbf{x}, t) = 0$$
$$\Psi^\dagger(\mathbf{x}, t)\Psi^\dagger(\mathbf{x}', t') - \Psi^\dagger(\mathbf{x}', t')\Psi^\dagger(\mathbf{x}, t) = 0 \qquad (8\text{-}31')$$

after a similar argument using the fact that $\partial\Psi(t)/\partial t$ and $\psi(t)$, $\partial\Psi^\dagger(t)/\partial t$ and $\Psi^\dagger(t)$ commute. Thus we have obtained all the commutation relations that satisfy relativistic requirements.

Even for more complicated tensor or spinor fields, the preceding argument may be applied as long as the fields can be expressed in canonical form and their components satisfy the Klein-Gordon equation; thus we can derive relativistic commutation relations from the canonical commutation relations. (I might remind you that the fact that the component satisfies the Klein-Gordon equation means that the field satisfies the de Broglie–Einstein relation but not necessarily that the field equation is the Klein-Gordon equation. The Dirac equation is a good example.) For cases other than the scalar field, however, the left-hand side of the commutation relation is not a scalar but in general a tensor or a spinor. Accordingly, the right-hand side cannot be a scalar like the Δ-function we just discussed. However, even for such cases, there always appears on the right-hand side some term in which some differential operators (including the zeroth differentiation), having the same transformation property as the left-hand side, are operated on Δ. For example, in the vector field, the field quantities are described by the four-vector $\Psi_1, \Psi_2, \Psi_3, \Psi_0$, and their commutation relation is

$$
\Psi_\mu(\mathbf{x}, t)\Psi_\nu^\dagger(\mathbf{x}', t') - \Psi_\nu^\dagger(\mathbf{x}', t')\Psi_\mu(\mathbf{x}, t) =
$$
$$
\frac{i}{\hbar}\left(g_{\mu\nu} + \frac{1}{\kappa^2}\nabla_\mu^0\nabla_\nu^0\right)\Delta(\mathbf{x} - \mathbf{x}', t - t'), \qquad \mu, \nu = 1, 2, 3, 0. \quad (8\text{-}32)
$$

Here, ∇_μ on the right-hand side is the μ-component of the differential operator ∇^0, defined in (8-16), and $g_{\mu\nu}$ is the $\mu\nu$-component of the metric tensor. Note that the left-hand side is of the form vector × vector, and therefore it is a second-rank tensor, but the right-hand side is also a second-rank tensor. In addition, we have relations that $\Psi_\mu(\mathbf{x}, t)$ and $\Psi_\nu(\mathbf{x}', t')$ commute and that $\Psi_\mu^\dagger(\mathbf{x}, t)$ and $\Psi_\nu^\dagger(\mathbf{x}', t')$ commute. I do not give an example here, but a similar situation exists for a spinor field.

Thus in the relativistic commutation relation derived from the canonical commutation relation, regardless of whether for a tensor or a spinor field, not only do the left- and right-hand sides have the same transformation properties, but also the right-hand side must always be Δ to which some covariant differential operator has been applied. This last point is essential; because of this, all commutation relations have identical form in all Lorentz systems. On the other hand, we can prove that there does not exist any invariant scalar function other than Δ which can be used. [If we abandon the initial conditions (II) and (III) which require the canonical commutation relation at $t = t'$, then there exists one more invariant scalar function Δ_1 in addition to Δ. However, this function does not have property (C), and therefore physical quantities at different points at $t' = t$ do not commute. From this follows the vexing situation that two different physical quantities at two different points cannot be measured simultaneously, and therefore Δ_1 is useless.]

Because of this situation, the commutation relation can *only* have the form

$$\Psi_\mu(\mathbf{x}, t)\Psi_\nu^\dagger(\mathbf{x}', t') - \Psi_\nu^\dagger(\mathbf{x}', t')\Psi_\mu(\mathbf{x}, t) = i\alpha D_{\mu\nu} \cdot \Delta(\mathbf{x} - \mathbf{x}', t - t') \quad \text{(8-33)}_-$$

if it is to satisfy the requirement that it retain its form in all Lorentz systems for any field. (This includes the case when the field is not canonically describable and we cannot start from canonical commutation relations.) α on the right-hand side is a real constant which cannot be determined from the aforementioned requirement alone, and $D_{\mu\nu}$ is a differential operator which has the same covariance property as the left-hand side. Furthermore, for the anticommutation relation, for which the relation with the canonical theory is not as natural, we also see that

$$\Psi_\mu(\mathbf{x}, t)\Psi_\nu^\dagger(\mathbf{x}', t') + \Psi_\nu^\dagger(\mathbf{x}', t')\Psi_\mu(\mathbf{x}, t) = i\alpha D_{\mu\nu} \cdot \Delta(\mathbf{x} - \mathbf{x}', t - t') \quad \text{(8-33)}_+$$

is the only relation satisfying the requirement that it have identical form in all Lorentz systems. [You might worry that for anticommutation relations, $\Psi(\mathbf{x}, t)$ and $\Psi(\mathbf{x}', t)$, and $\Psi^\dagger(\mathbf{x}, t)$ and $\Psi^\dagger(\mathbf{x}', t)$ do not commute and therefore all the field quantities at different places cannot be simultaneously measured; I shall address this concern shortly.]

In the discussion given so far, it did not matter whether the field is a tensor or a spinor, but now the difference emerges. For a tensor field the left-hand side of the commutation relation is always an even-rank tensor. Therefore, $D_{\mu\nu}$ must be an even-order differential operator. Thus if we consider

$$\begin{aligned}
X \equiv & [\Psi_\mu(\mathbf{x}, t)\Psi_\nu^\dagger(\mathbf{x}', t') \mp \Psi_n^\dagger(\mathbf{x}', t')\Psi_\mu(\mathbf{x}, t)] + \\
& [\Psi_\mu(\mathbf{x}', t')\Psi_\nu^\dagger(\mathbf{x}, t) \mp \Psi_\nu^\dagger(\mathbf{x}, t)\Psi_\mu(\mathbf{x}', t')] \qquad \text{(8-34)} \\
= & i\alpha\{D_{\mu\nu} \cdot \Delta(\mathbf{x} - \mathbf{x}', t - t') \\
& + [\text{the same as on the left except that } (\mathbf{x}, t) \text{ and } (\mathbf{x}', t') \\
& \quad \text{are interchanged}] \},
\end{aligned}$$

since $D_{\mu\nu}$ on the right-hand side is differentiation of even order, the symmetry property of a function does not change when it is applied. Therefore, the right-hand side should have the same symmetry property as

$$\Delta(\mathbf{x} - \mathbf{x}', t - t') + \Delta(\mathbf{x}' - \mathbf{x}, t' - t).$$

In other words, due to (B) given earlier, it should be symmetric with respect to interchange of $\mathbf{x}$ and $\mathbf{x}'$ and antisymmetric with respect to interchange of t and t'. On the other hand, since the left-hand side is symmetric with respect to interchange of $(\mathbf{x}, t)$ *and* $(\mathbf{x}', t')$, if it is symmetric with respect to interchange of $\mathbf{x}$ and $\mathbf{x}'$, then it must also be symmetric for interchange of t and t'. Therefore, we are led to conclude that the symmetry properties of the right- and left-hand sides are different. From this we realize we must have

$$X(\mathbf{x} - \mathbf{x}', t - t') = 0.$$

This does not pose any problem if we use the minus sign in the top expression of (8-34). Indeed that was the case for the Klein-Gordon field. However, a contradiction results for the plus sign. For if we put $\mathbf{x} = \mathbf{x}', t = t', \mu = \nu$, we obtain

$$X = 2\left[\Psi_\mu(\mathbf{x}, t)\Psi_\mu^\dagger(\mathbf{x}, t) + \Psi_\mu^\dagger(\mathbf{x}, t)\Psi_\mu(\mathbf{x}, t)\right] = 0,$$

but since the eigenvalues of neither $\Psi\Psi^\dagger$ nor $\Psi^\dagger\Psi$ can ever be negative, $X = 0$ means $\Psi = 0$, $\Psi^\dagger = 0$. Thus we arrive at the conclusion that it is impossible to use anticommutation relations for a tensor field. This means that the particle associated with a tensor field must be a boson. So far we have been discussing cases in which the field is described by complex Ψ and $\Psi^\dagger$, but we arrive at the same conclusion for real Ψ.

What happens for a spinor field? Again our lack of knowledge of spinor algebra hampers our argument, but we can say that in this case $D_{\mu\nu}$ is a differentiation of odd order. In this case, X in (8-34) does not have to be zero, and therefore we do not have to worry about the problem of $\Psi = \Psi^\dagger = 0$, even if the anticommutation relation is used. Therefore, for this case, either the anticommutation relation or the commutation relation can be valid without contradiction. However, as I pointed out earlier in (8-13'), negative energy results for a spinor field, and thus to avoid this difficulty, the particle must be a fermion, and therefore the anticommutation relation must be used.

If we summarize all these conclusions, we obtain (8-14'). Here it is very significant that the anticommutation relation is possible only for a spinor field. I told you earlier that for the case of an anticommutation relation, Ψ and $\Psi^\dagger$ at different points at the same time do not commute, and therefore we cannot measure the field at different points simultaneously. However, in this case, Ψ and $\Psi^\dagger$ are spinors, and as I told you in the previous lecture, what is physically meaningful is not Ψ or $\Psi^\dagger$ itself but tensors composed of an even number of Ψ or $\Psi^\dagger$. It should be noted that these tensors do not change sign upon a 2π rotation of the coordinate axes and they always commute for different points at the same time.

Now I am going to end this long story, but before claiming the validity of the title of this lecture, there is still a gap in the logic. Namely, we have not yet proved that tensor fields correspond to integral spin and spinor fields to half-integral spin. In order to do that, we need not only spinor algebra but also the complicated mathematics of angular momentum, and to my regret I cannot go into it. However, I can roughly say as follows. We can prove the above statement using the facts that (a) addition of an even number of 1/2 spins gives only integral angular momentum and addition of an odd number of 1/2 spins gives only half-integral angular momentum and (b) an even-rank spinor can be regarded as a tensor, but an odd-rank spinor is still a spinor.

Anyhow, it is extremely interesting that the important results of the relation between spin and statistics can be derived from the most fundamental requirements only—such as the covariance with respect to Lorentz transformation and the de Broglie–Einstein relation. In concluding this long story, let me quote Pauli's words at the end of his paper. "We wish to state, that according to our opinion the connection between spin and statistics is one of the most important applications of the special relativity theory."

As I told you at the beginning of today's lecture, Pauli published this paper in 1940, but I might note that the four-dimensional commutation relation that Pauli used there had already been discussed in 1927 in his paper with Jordan on the electromagnetic field. It was already pointed out in this paper that the right-hand side and the left-hand side of the four-dimensional commutation relations for $\mathbf{E}(\mathbf{x}, t)$ and $\mathbf{H}(\mathbf{x}, t)$ should have the same covariance properties and that the right-hand side is a covariant differential operator applied to an invariant scalar function Δ.

Finally, I would like to point out that the relation between spin and statistics discussed so far applies to elementary particles. But exactly the same conclusions are obtained for composite particles (for example, for atoms and molecules or general atomic nuclei). I shall talk about this in the next lecture.

The Year of Discovery: 1932

The Discovery of the Neutron and Ensuing New Developments

Since I have talked quite a bit using mathematics, today I shall talk without mathematics. When I discussed the spin and statistics of elementary particles, I mentioned some events that formed the background for Pauli's study of this problem. They were the discovery of the neutron in 1932 and several theoretical developments (Heisenberg's theory of nuclear structure, Fermi's theory of β decay, and Yukawa's meson theory) that were stimulated by it; again on the experimental side was the discovery of a new particle by Anderson and Neddermeyer. Today we shall return to 1932 and talk in somewhat greater detail about the happenings around that time.

As I told you last time, 1932 was truly quite a year. It was the year in which Anderson discovered the positron and Chadwick discovered the neutron. In addition, the chemist H. Urey discovered hydrogen with atomic weight 2, the so-called heavy hydrogen (deuterium), and J. Cockcroft and E. Walton finished building their ingenious apparatus for accelerating particles, with which they destroyed Li nuclei using protons as bullets. Moreover, all these related discoveries appeared with such fantastic timing that they stimulated one another with a tremendous multiplicative effect. For example, the spin and magnetic moment of the neutron, which are difficult to measure directly, were determined indirectly by measuring those of the deuterium nucleus.

As for every discovery, those mentioned above involved a succession of many episodes. Some scientists were very close to these discoveries but missed them, and other scientists captured what other people missed by adopting methods which with hindsight were obvious, like Columbus' egg.[1] Also, some scientists

---

1. "Columbus' egg" is an idiom which seems to be known in Eastern Europe and Asia, but not in North America. The apocryphal origin of the idiom is the following. When someone told Christopher Columbus that anybody could have discovered the New World, Columbus challenged

Figure 9.1 James Chadwick (1891–1974).
[Photograph by Börtzells Esselte. Courtesy
of AIP Emilio Segrè Visual Archives]

serendipitously discovered things. For example, Anderson was not attempting
to discover the positron when he did. He was trying to determine the energy of
charged particles in cosmic rays and was taking photographs of particle paths
in a Wilson cloud chamber under a strong magnetic field when he found the
flight tracks of positrons in many pictures.

Many scientists beside Anderson were doing similar experiments. Why
did they fail to discover the positron? Although one reason was that their
experiments were not as large scale as Anderson's, the chief reason was that
they did not try to establish in which direction the particles were traveling.
Anderson examined the directions of particles merely by inserting a lead plate
into the cloud chamber. In passing through the plate, a particle might lose energy
but never gain energy. Therefore, if you measure the radius of curvature of the
track on both sides of the plate and examine on which side the particle had
greater energy, then you can definitively decide from which side the particle hit
the plate. If the direction is thus determined, we can determine the sign of the
charge by observing whether the trace curved right or left.

---

him to stand a raw egg upright on its end on a table. After no one in the room could accomplish
this, Columbus took the egg, crushed one end of it and then stood it upright on the table. (The
reader may wish to try this; it does work!)

There is a more complicated and stranger-than-fiction story on the discovery of the neutron. Although this story is not directly related to spin, let me talk about it because it is so interesting.

The story goes back to around 1930. At that time W. Bothe and H. Becker were studying γ-rays by bombarding many atomic nuclei with α-rays from polonium. After a series of experiments, they found that beryllium nuclei emitted extremely strong "γ-rays." (Bothe and Becker were thinking that what came out was a γ-ray just as from other nuclear species, but this was not completely certain, so I write "γ-ray" with quotation marks. It was shown later that a γ-ray was indeed coming out of Be, but something else was also coming out.) Furthermore, this "γ-ray" has the characteristic of extremely strong penetration, and this aroused many people's attention.

Among the people who became interested in this phenomenon were I. Curie and F. Joliot. These two scientists, having the advantage of being Curies, repeated Bothe and Becker's experiment using an otherwise unobtainably large amount of Po and discovered a peculiar thing.

In order to examine the absorption of this "γ-ray" in passing through various materials, the Curies put plates of all sorts of materials at the window of the ionization chamber (an apparatus which measures the intensity of radioactive radiation by measuring the amount of ions created in the gas) in order to obtain the effect of the "γ-rays" entering the ionization chamber. From this experiment the Curies found that the presence or absence of Pb, Cu, or C plates had almost no effect on their measured values, but they observed extraordinarily many ions in the ionization chamber for samples containing many hydrogens, such as water or paraffin. Such a phenomenon had never before been observed for γ-rays.

The Curies arrived at the following conclusion after doing many more experiments to understand this unexpected phenomenon. You must know that when an X-ray or γ-ray passes through material, its $h\nu$ scatters electrons from atoms. These are scattered electrons observed in the Compton effect. The Curies thought that if $h\nu$ became very large, then it would scatter not only electrons but also light nuclei such as protons. They confirmed by calculation that such a phenomenon was indeed possible if $h\nu$ of the γ-ray was on the order of 50 MeV. Therefore, if a γ-ray with a large $h\nu$ went through water or paraffin, many protons would be scattered and then go into the ionization chamber. Then the gas in the chamber would be ionized more strongly by those protons than by γ-rays, for a charged particle has a higher ionization capability than a γ-ray. Thus the Curies discovered that the "γ-ray" from Be scattered many protons from water and paraffin. This phenomenon was discovered in 1931.

Needless to say, this information was immediately transferred to Chadwick. Like the Curies, he was interested in the "γ-ray" of Bothe and Becker and was trying many experiments. He was especially interested by the proton scattering

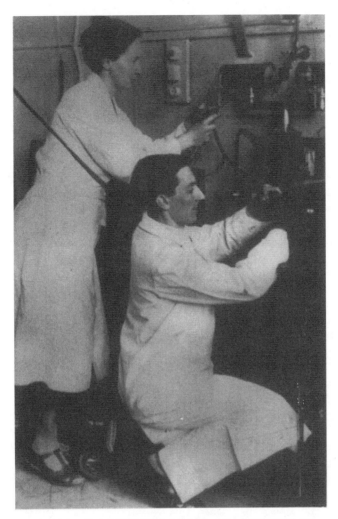

Figure 9.2 Irène Curie (1897–1956), J. Frédéric Joliot (1900–1958). [Photograph by Société Française de Physique, Paris. Courtesy of AIP Emilio Segrè Visual Archives]

found by the Curies and reexamined the phenomenon with a more elaborate procedure than they had used. He observed that the scattered protons indeed existed and were flying with a speed on the order of 3×10^9 cm s^{-1}. He therefore confirmed the Curies' conclusion that the scattered protons were coming from water and paraffin.

However, Chadwick doubted the Curies' interpretation of the phenomenon being due to γ-rays. He proceeded as follows. If indeed the scattered protons with the velocity of 3×10^9 cm s^{-1} were produced by $h\nu$ of 50 MeV, then if the same $h\nu$ scattered a nitrogen nucleus from a nitrogen atom, what would be the speed of the scattered nitrogen nucleus, and how many ions would the nitrogen nucleus produce in the ionization chamber? Chadwick calculated this and checked it with an experiment using nitrogen. He found that many more ions came out in the experiment than the calculation predicted. Furthermore, the other fatal flaw that he pointed out in the Curies' interpretation was that $h\nu$ with this tremendous energy of 50 MeV was very unlikely to be emitted from Be. Namely, he knew the energy of the α-particle from Po, the mc^2 of Be, and the mc^2 of C, which is produced when an α-particle combines with Be. Then the magnitude of the $h\nu$ coming from carbon can be calculated by energy balance, and $h\nu$ is at most 10 MeV.

Therefore Chadwick, unlike the Curies, thought that the "γ-ray" coming from Be was not a γ-ray and suspected that it was a particle without charge and with a mass nearly equal to that of the proton. Chadwick reasoned this way because, if the mass of the particle is comparable to that of the proton, then the particle can scatter a proton without having the outrageously large energy of 50 MeV. You must have learned that in the Compton effect only a small fraction of the $h\nu$ of the photon is transferred to the scattered electron. This also applies to the case of a scattered proton. It was for this reason that the large energy of 50 MeV was needed for $h\nu$. However, if the incoming particle has a mass similar to that of the proton as Chadwick considered, then it can transfer its entire energy to the scattered proton during a collision. (If you slide a nickel squarely into another nickel, the hitting nickel will stop, and the hit nickel will receive all the energy.) Chadwick further did a variety of calculations and experiments (among them was the problem of scattered nitrogen, which I mentioned earlier) and accumulated other circumstantial evidence that what was coming from Be was a neutral particle with the same mass as the proton and decided to publish the discovery of a new particle in February 1932. This particle was named *neutron*.

Thus even hotshots like Curie and Joliot just missed discovering the neutron. They missed it perhaps because they had a preconception that the "γ-ray" was a γ-ray. On the other hand, Chadwick, as he himself mentioned in his Nobel lecture, was influenced by his mentor Rutherford, who anticipated around 1920 the existence of neutral particles with mass similar to that of the proton. Therefore, it is certain that such an idea was rooted in Chadwick's mind. Since he quotes Rutherford's words in his lecture, let me repeat them here.

"Under some conditions it may be possible for an electron to combine much more closely with a hydrogen nucleus, forming a kind of neutral doublet . . . Its external field would be practically zero . . . and in consequence it should be able to move freely through matter."

Chadwick, who was faithful to his mentor, not only took these words of his mentor to heart but also tried all sorts of experiments to actually discover this neutral particle. Therefore, I imagine that as soon as the Curies detected the scattered proton, Chadwick must have immediately remembered the neutral particle mentioned by Rutherford.

Actually, there was another scientist before Chadwick who did an experiment believing that the "γ-ray" might be a neutral particle. His name was H. C. Webster. He seems to have made an unfortunate choice of how to do the experiment, and he did not discover the neutron.

The first task after the discovery of the neutron is to determine its accurate mass and its spin, statistics, and magnetic moment. Now since the neutron is neutral, we cannot use the methods applicable to nuclei. However, it was lucky that, as I told you earlier, the deuteron was also discovered at this time. Because a deuteron is a combination of a proton and a neutron, quite a bit of information about the neutron was indirectly obtained from the deuteron. Let me now go into this story.

Since the mass number of the deuteron is 2 and its charge is 1, we can imagine that it is a combination of a neutron and a proton. This was experimentally confirmed in 1934, two years after the discovery of the neutron, when Chadwick and M. Goldhaber bombarded the deuteron with γ-rays and dissociated it into a proton and a neutron. This experiment determined the mass of the neutron. The masses of the proton and deuteron were already known, the energy of the γ-ray used in the experiment was also known, and the kinetic energy of the proton from the decomposition could also be measured. Therefore, by combining these data, the mass of the neutron could be exactly calculated from the balances of energy and momentum. Thus 1.0085 was obtained. The mass of the proton is 1.00807, and thus the neutron is slightly heavier than the proton.

Next, as I said earlier, it is difficult to determine the neutron's spin and statistics directly, and therefore we must determine them from the spin and statistics of the deuteron. The spin and statistics of the deuteron were obtained in 1934 through the analysis of the band spectrum of the deuterium molecule (remember lecture 4). This work was done by G. M. Murphy and H. Johnston; they found that the deuteron obeys Bose statistics and its spin is 1. Now that we know the spin of the deuteron, the spin of the neutron must be half-integral, such as $1/2, 3/2$, from the addition of angular momenta (lecture 5) and from the fact that the proton spin is $1/2$. As for statistics, since the deuteron obeys Bose statistics and the proton is a fermion, it is concluded that the neutron must be a fermion according to Ehrenfest and Oppenheimer's rule (in reverse), which I discuss below.

This last conclusion is noteworthy. Because Rutherford's neutron was a combination of a proton and an electron, its spin would have to be an integer and its statistics according to the Ehrenfest-Oppenheimer rule would be Bose statistics. Therefore, the neutron discovered by Chadwick was different from that conceived by Rutherford. Chadwick was faithful to his mentor, but the neutron itself was not faithful to the great prophets who had predicted its existence. By the way, the probability that the neutron spin could be 3/2 has been excluded by various other points of view.

The magnetic moment of the neutron is also difficult to determine directly; the only way to determine it was by calculating it from those of the proton and deuteron. It may sound incredible that not only the magnetic moment of the newly discovered deuteron but also the magnetic moment of the long-known proton were not yet measured in 1932. Nevertheless, this is a fact. The proton magnetic moment was first measured by O. Stern and I. Estermann in 1933. As I told you at the end of lecture 4, Stern measured it by his favorite technique of bending a molecular beam by an inhomogeneous magnetic field and observed that the proton magnetic moment is 2.5 $e\hbar/2m_p c$. (Here m_p is the mass of the proton, and just as we say $e\hbar/2mc$ is a Bohr magneton, we call $e\hbar/2m_p c$ a nuclear magneton.) Rabi and his collaborators did a more accurate experiment in 1934 and obtained 2.79 $e\hbar/2m_p c$, and they simultaneously measured the magnetic moment of the deuteron to be 0.86 $e\hbar/2m_p c$. Furthermore, they confirmed that the direction of the magnetic moment is the same as the direction of spin for both the proton and the deuteron.

If we thus know the magnitude and the direction of the magnetic moments of the proton and the deuteron, then we can calculate the magnetic moment of the neutron from the relation

$$\mu_{deuteron} = \mu_{proton} + \mu_{neutron} \tag{9-1}$$

(here we wrote the magnetic moment in units of the nuclear magneton as μ). We then obtain $\mu_{neutron} = -1.93$. The negative sign shows that the direction of the magnetic moment is opposite to that of spin. In doing this we are implicitly assuming that the spin 1/2 of the proton and the spin 1/2 of the neutron are parallel to each other, making the spin of the deuteron 1. In other words, we are assuming in the deuteron that the orbital angular momenta of its component proton and neutron are equal to 0. That is to say, we are assuming that the deuteron is in an S state. This assumption is justified theoretically. Since the binding energy of the deuteron is extremely small, we can theoretically exclude the possibility that a proton and a neutron can combine into a state other than an S state.[2]

---

2. It was shown by Schwinger that the ground state of deuterium is a mixture of ^3S and ^3D states and fully described in Rarita W and Schwinger J 1941 On the Neutron-Proton Interaction *Phys. Rev.* **59** 677–695.

Since the proton is a fermion with spin $1/2$, many people had thought, until its magnetic moment was measured, that it should obey the Dirac equation. If that were the case, its magnetic moment would have been $e\hbar/2m_pc$, but in fact the measured value was 2.79 times as large as that. We say that the proton has an *anomalous magnetic moment*. The neutron, likewise, has an anomalous magnetic moment, which was first explained, though qualitatively, in the context of Yukawa's meson theory. I shall talk about it later; here I just narrate an episode on this anomaly.

I told you that Estermann and Stern measured the magnetic moment of the proton. One day Pauli visited Stern's laboratory. Pauli asked Stern what he was doing. Stern told him that he was measuring the magnetic moment of the proton. Then Pauli said it was meaningless to do such an experiment. "Don't you know the Dirac theory? It is obvious from Dirac's equation that the moment must be $e\hbar/2m_pc$." I guess the impact that the Dirac equation made on Pauli was quite stunning. (This gossip originated from J. H. D. Jensen when he came to Japan and visited the University of Tokyo the year before last. I did not listen to him directly, but this story was relayed to me by Professor Fujita of the Tokyo University of Education.)

When the neutron was discovered, the idea that an atomic nucleus is composed of protons and neutrons occurred to several people. Among them was Heisenberg, who, as soon as he heard news of Chadwick's discovery, started his epoch-making work, which gives a very clear explanation of nuclear properties based on this idea. In 1932 he published three papers discussing the structure of nuclei based on this idea; they all abound in ideas and stimulated many people. Before going into this subject, it is necessary to know what people had thought before the discovery of the neutron.

Before the neutron was discovered, the idea was prevalent that nuclei were composed of protons and electrons. As evidence of this, we can quote the paper by Ehrenfest and Oppenheimer. These two distinguished scientists wrote a paper in 1930 titled "Note on the Statistics of Nuclei" and theoretically proved that the following rule should be valid for the statistics of nuclei: If we have two nuclei each built up from n electrons and m protons, and if $n + m$ is even (odd), then only states whose total wave function does not change (does change) its sign upon interchange of nuclear coordinates are actually allowed. (Read this remembering diatomic molecules, which were discussed in lecture 4.)

If you read this sentence, it is obvious that they tentatively assumed that a nucleus is composed of protons and electrons. The essential point in their proof of this rule is as follows. Both the electron and the proton are fermions. Therefore, each time the coordinates of two electrons or two protons are interchanged, the total wave function changes its sign. On the other hand, interchange of the nuclear coordinates means interchange of electrons n times and interchange of protons m times, and therefore if $n + m$ is even (odd), then the sign of the wave function changes an even (odd) number of times and therefore only those states whose wave function does not (does) change sign with the interchange of nuclear coordinates are allowed. This is the central point in their proof. Remembering lecture 4 on the relation between the symmetry of wave functions and statistics, we can rephrase the rule as "If $n+m$ is even, the nucleus obeys Bose statistics, and if it is odd, Fermi statistics." (Actually, what I just described as the essential point of this proof is almost self-evident, and it was not worthy for these two distinguished scientists to have written a paper about it. It looks like what they wanted to do here was to prove the rule much more rigorously and to examine its limit carefully. They concluded that if two nuclei are very close or interacting so strongly that we cannot regard them as separately closed systems, then this rule cannot be applied literally. Furthermore, they do not assume anywhere in their proof that nuclei are composed of protons and electrons, so the rule applies as long as the nuclei are composed of fermions. More generally, there do not have to be only two species, but any closed particle system composed of an arbitrary number of fermions and an arbitrary number of bosons obeys Bose statistics (Fermi statistics) if the number of fermions is even (odd), and the reverse of this rule is also true.)

What will happen if we apply this rule to actual nuclei? Let us find out.

If a nucleus is composed of n electrons and m protons, then the mass number A and charge Z are given by

$$A = m, \qquad Z = m - n, \tag{9-2}$$

which means

$$m + n = 2A - Z, \tag{9-3}$$

and therefore we can rephrase the Ehrenfest-Oppenheimer rule by saying that a nucleus obeys Bose (Fermi) statistics if $2A - Z$ is even (odd). On the other hand, as I told you in lecture 4, it is possible to experimentally determine the nuclear statistics and spin from experimental band spectra. Therefore, we can experimentally check the validity of this rule, given in final form above.

If we do this for the bands of H_2, O_2, and He_2, we find that the H nucleus obeys Fermi statistics, the O nucleus obeys Bose statistics, and the He nucleus obeys Bose statistics, and thus all of them obey this rule, and everything is OK. However, if we examine the band of N_2 experimentally, we obtain the conclusion that the N nucleus obeys Bose statistics. However, for the N nucleus, $Z = 7, A = 14$, and therefore $2A - Z = 21$ is odd, and if we apply the Ehrenfest-Oppenheimer rule, we conclude that the N nucleus obeys Fermi statistics, and therefore this rule is violated. (I might also point out that for the D nucleus, $A = 2, Z = 1$, and therefore $2A - Z = 3$ is odd. According to the rule, it should obey Fermi statistics. Nevertheless, the experiment by Murphy and Johnston shows that it obeys Bose statistics.)

This difficulty of the N nucleus had been known since 1929.[3] Ehrenfest and Oppenheimer knew it very well, of course, and in fact they explicitly point that out in their paper. Since they knew this example which violated their rule, what were they thinking about this violation when they published their paper? It is not quite clear whether they thought that the prevailing theory that a nucleus was composed of an electron and a proton should be abandoned or that the theory must be accepted for the time being. In this connection perhaps we should remember that around 1930 Bohr and quite a few other luminaries were close to believing that quantum mechanics did not work in nuclei.

Therefore, if Ehrenfest and Oppenheimer were influenced by this belief, then they would not have concluded that the idea that a nucleus was composed of electrons and protons should be abandoned simply because there is an instance in which their rule was violated. We can take the view that the contradiction posed by the N nucleus is evidence that the quantum mechanics they used for deriving the rule is not operating in the nucleus. Regardless of whether they approved or disapproved of the idea that nucleus = n electrons + m protons, perhaps they presented their rule as they did because of the views prevailing at that time.

There is one more phenomenon that was blamed on the premise that quantum mechanics did not work in the nucleus. This is the fact that the β-rays emitted in β-disintegration have a continuous spectrum. Suppose nucleus A is transmuted to nucleus B by emitting a β-ray. Here the masses of A and B are definite. Therefore, because of energy conservation, the difference of their mc^2 must be the energy of the β-ray. If this is the case, the energy of the β-ray must have a definite, discrete value, and therefore a line spectrum should be observed. In reality, however, the spectrum is continuous. Bohr asserted that this difficulty appeared because quantum mechanics did not work in the nucleus, and therefore the law of energy conservation was violated. Bohr said that since an electron with a Compton wavelength of 10^{-11} cm was confined in the small nucleus with a 10^{-13} cm radius, the individuality of the electron was lost in the nucleus. (A scientist of Bohr's stature did not say it in such a childish way. He seems to have been considering the Klein paradox.)[4] The bottom line is that according to

3. This was noted by W. Heitler and G. Herzberg (1929 Gehorchen die Stickstoffkerne der Boseschen Statistik? *Naturwissenschaften* **17** 673–674) in the rotational Raman spectrum of N_2 observed by F. Rasetti (1929 On the Raman Effect in Diatomic Gases. II *Proc. Natl Acad. Sci. USA* **15** 515–519).

4. "Klein announced [Klein O 1929 Die Reflexion von Elektronen an einem Potentialsprung nach der relativistischen Dynamik von Dirac Z. *Phys.* **53** 157–165] his paradox which caused such headaches to Bohr and others. Klein had noted that the Dirac equation then leads to absurdities when one attempts to localize electrons with a precision $\leq \hbar/mc$." Pais, *Inward Bound*, p. 349.

Bohr's idea, the interior of a nucleus was a sanctuary that could not be penetrated by quantum mechanics.

This does not mean that there was no opposition to the theory of a nuclear sanctuary around 1930. For example, Pauli had an idea, which goes as follows. There exist in the nucleus particles called "neutrons" which are electrically neutral with a mass of an electron or nearly zero, and when the nucleus undergoes β-disintegration, an electron is ejected together with this "neutron." If we so hypothesize, the *sum* of the energies of the electron and this neutral particle is definitely due to the energy conservation law, but the energy of the electron itself or the energy of the neutral particle itself can have a continuous spectrum. This was Pauli's idea. Pauli thought this same idea could solve the problem of the N nucleus. Apparently he talked about it to Bohr, but it seems he could not convince Bohr. Not only that, he could not convince his close friend Heisenberg.

Either because of his well-known perfectionist tendency or because of his hesitation due to the fact that such a particle had never been observed, Pauli did not publish this idea in a paper but proposed it to his friends in letters around 1930. The original is in German, but I have an English translation here.

"I came to a desperate conclusion . . . that inside the nucleus there may exist an electrically neutral particle which I shall call *neutron*. The continuous β-spectrum is understandable if one assumes that, during β-decay, the emission of an electron is accompanied by the emission of a neutron."

Since he wrote this in 1930, it was before Chadwick discovered the neutron. However, Pauli's "neutron" is different from Chadwick's neutron because of its mass. Pauli's idea was adopted four years later by Fermi, who solved the problem of the β-ray; in order to discriminate Pauli's neutron and Chadwick's neutron, Fermi called the former a "neutrino."

The fact that Pauli did not publicize his idea and the fact that he writes in his letter "desperate conclusion" show how timid people were at the time in considering the existence of new particles other than the proton and the electron. As I told you earlier, Rutherford was considering the existence of the neutron around 1920, but even that was considered to be a combined proton and electron rather than a new particle.

However, spurred on by Chadwick's discovery of the neutron, the situation began to change drastically. As I told you earlier, as soon as he heard of Chadwick's discovery, Heisenberg proposed that it is not necessary to assume the existence of electrons in the nucleus but rather that the nucleus is composed of protons and neutrons. He showed that, based on this idea, he could explain most of the intranuclear problems quantum-mechanically, and he destroyed for the first time the idea of a nuclear sanctuary. However, he retained the phenomena related to β-rays within the sanctuary and did not lay his hands on them. It was Fermi who destroyed this reserved idea of Heisenberg's when in 1934 he treated β-decay quantum-mechanically, exploiting the idea of Pauli's neutrino, which nobody had adopted until then. However, the other problem in the sanctuary that Heisenberg left untouched, i.e., the origin of the exchange force between protons and neutrons, which plays an important role in his nuclear theory, was not addressed by anybody for some time despite Fermi's success. It was none other than Yukawa who opened up this topic by introducing a new particle, called *meson*, and invaded the sanctuary with quantum mechanics.

Thus the history of physics after the discovery of the neutron in 1932 became a history of pulling down one after the other the walls of the nuclear sanctuary, which had been thought to be impenetrable by quantum mechanics.

I was thinking of talking to the bitter end of this historical development, but I am running out of time. Since this is the last climax remaining before the Second World War, I would like to postpone any further discussion and spend ample time on it in the next lecture rather than do it now in a hurry. In this process, the important concept of *isospin* was born. This is different from the spin associated with angular momentum, but mathematically it has very similar properties and plays a major role in elementary particle theory. I will spend ample time on it in the next lecture.

LECTURE TEN

Nuclear Force and Isospin

From Heisenberg to Fermi and from
Fermi to Yukawa

As I promised you last time, I shall talk about how the walls of the sanctuary, the interior of the nuclei, were gradually removed first by Heisenberg's theory of nuclear structure and further by Fermi's and Yukawa's. In this process I shall explain how the new concept of *isospin*,[1] which is the sibling of spin, was born. This concept came to play an essential role later in nuclear theory and in elementary particle theory.

Heisenberg inaugurated his theory of nuclear structure when he submitted his first paper on it to *Zeitschrift für Physik* in June 1932. It was in February of the same year that Chadwick submitted the report of his discovery of the neutron to *Nature*, and this shows how quickly and efficiently Heisenberg worked. This paper was epoch making not only because it contains the simple idea that atomic nuclei are composed of neutrons and protons but also because it proposes that the force acting between neutrons and protons is an exchange force, and in order to describe this force, the important new concept of *isospin*, just mentioned, was introduced. Heisenberg finished writing the second paper in the next month, July, and the third paper in December and gave a truly clear theoretical explanation of a variety of nuclear properties. Thus the first breach was made in the walls of the sanctuary which had so far been considered unassailable by quantum mechanics.

The start of his paper can be paraphrased as follows:

> From the research of the wife-and-husband team Curie and Joliot and its interpretation by Chadwick, we have learned that the new particle named neutron plays a role in the atomic nucleus. This discovery suggests

1. Throughout the text Tomonaga uses the Japanese phrase which translates as *charge spin;* I have replaced it with the English word *isospin.*

162

Figure 10.1 Heisenberg, 1931.
[Photograph by Max-Planck-
Institut. Courtesy of AIP Emilio
Segrè Visual Archives]

that an atomic nucleus is composed of neutrons and protons and that the electron is not one of its constituents. If this hypothesis is correct, then the theory of the atomic nucleus is greatly simplified. The fundamental problems which have bothered people so far, such as why the β-ray has a continuous spectrum and why the nitrogen nucleus obeys Bose statistics, are reduced to more fundamental problems such as what is the rule when a neutron disintegrates into a proton and an electron, why does the neutron obey Fermi statistics, etc. [right from the outset, he is assuming that the neutron is a fermion]; the problem of nuclear structure itself could be separated from these profound problems, and we can use quantum mechanics to explain them as the result of the force between neutrons and protons, etc.

This is the basis of Heisenberg's idea.

As I pointed out to you before, his first requirement in this paper is that the neutron be a fermion with spin 1/2. This requirement was necessary to explain the experimental results on the statistics and spin of the nitrogen nucleus. (Murphy's experiment had not yet been done, and nothing was known about the statistics and spin of the neutron.) In fact as I noted earlier, the Ehrenfest-

Oppenheimer rule holds if the nucleus is composed of fermions, and therefore their rule holds if the neutron is a fermion by replacing their statement about electrons by one about neutrons; in applying this rule we can use $A = m+n$ and $Z = m$ instead of (9-2), and therefore the nitrogen nucleus with even A obeys Bose statistics. This agrees with the empirical rule from the band spectrum. As for spin, the addition rule of spin says that the total spin is an integer if A is even, and this also agrees with experiment.

For these reasons Heisenberg thought that the neutron is a fermion, but he does not venture on why it must be so. Anyhow, he says that the idea that a neutron is composed of a proton and an electron is not effective and furthermore considers the neutron as equally independent an elementary particle as the proton. But he assumes that, depending on the circumstance, a neutron may split into a proton and an electron, and in this splitting quantum mechanics probably does not apply, and the conservation of energy and the conservation of momentum are violated. In this last point, Heisenberg shared Bohr's view, which I mentioned in the last lecture, and is opposed to Pauli.

Therefore, Heisenberg spares the sanctuary in which conservation laws are violated and quantum mechanics cannot help, but he could explore quite a bit of nuclear physics without disturbing the sanctuary. This is Heisenberg's theory of nuclear structure.

In order to introduce quantum mechanics into the nucleus, we must know what type of force acts between the neutron and the proton, its constituents. Since this force strongly binds neutrons and protons into a clump the size of nuclei, it must be a very strong attraction. On the other hand, from the experiment of proton scattering by nuclei, it is unlikely that such a force extends over a long distance. Therefore, this force must act only when the particles approach within a distance on the order of the radius of nuclei (10^{-13} cm). Let us call this force the *nuclear force*. We can consider three kinds of forces, i.e., between neutron and neutron, proton and proton, and proton and neutron. Heisenberg reasoned as follows as to which of them plays the most important role.

It is an experimental fact that if the atomic number Z is not very large and therefore the charge Ze is not very large, then Z is approximately $A/2$ for many nuclei, where A is the mass number. This shows that nuclei with approximately equal numbers of neutrons and protons are most stable. From this fact Heisenberg concluded that the attraction between neutrons and protons plays the biggest role in the nucleus. For if the attraction between neutrons were stronger, then nuclei composed only of neutrons would be more stable, and therefore more such nuclei should exist. But this contradicts the facts. The same thing can be said if the attraction between protons is stronger—namely, nuclei with only protons must be abundant. (The story is different if Z is very large and Coulomb repulsion is very strong.) For these reasons he considered only the force between neutrons and protons for the time being.

Next Heisenberg noticed the experimental fact that the binding energies of nuclei are approximately proportional to the mass number A (i.e., the number of particles in the nucleus). From this he was led to the idea that the nuclear force is not the usual attractive force but is an exchange force. (I will soon explain the meaning of exchange force.) He reasoned as follows. If the force acting between a neutron and a proton is the usual two-body force, then if the potential between the Kth neutron and the Lth proton is written as $V_{K,L}$, and if the number of neutrons is written as N and the number of protons is written as P, then the total potential is

$$\sum_{K=1}^{N} \sum_{L=1}^{P} V_{K,L},$$

and the total binding energy must approximately equal the number of combinations of pairs (K,L), which is $N \cdot P \approx A^2/4$. In reality, it is proportional only to A.

From this fact Heisenberg cut to the essence of the nuclear force and thought it must be an *exchange force*. The concept of exchange force appears in the field of chemistry. For example, when two hydrogen atoms combine to form a hydrogen molecule, the force which binds the atoms is an exchange force. The special characteristic of this force is that once a hydrogen molecule is thus produced, even though a third hydrogen atom comes close, the hydrogen atoms in the molecule no longer attract the third atom. Therefore, if we assemble many hydrogen atoms to make a drop of liquid hydrogen, the binding energy of the drop is essentially only the sum of the binding energies of the hydrogen molecules. Therefore, it is only proportional to the number of hydrogen atoms in the droplet.

Now let me talk about the process occurring when two hydrogen atoms attract one another to combine into a hydrogen molecule and why we call the force responsible for it an exchange force. Since the case of H_2^+ ion is simpler than the case of hydrogen molecule and is closer to that of a neutron and a proton, let me use it as an example as we proceed.

The H_2^+ ion is a mechanical system composed of two protons and one electron. For simplicity let us assume that the two protons are fixed in space. We shall label these two nuclei as nucleus I and nucleus II. Let us first consider the case in which the two nuclei are infinitely far apart. We can consider two states. The first state is when nucleus I and the electron form a hydrogen atom and nucleus II is naked. The second is when nucleus I is naked and nucleus II and the electron form a hydrogen atom. Let us write the coordinate of the electron as $\mathbf{x}$ and the wave functions for these two states as $\psi_\mathrm{I}(\mathbf{x})$ and $\psi_\mathrm{II}(\mathbf{x})$. It is obvious that $\psi_\mathrm{I}(\mathbf{x})$ is the wave function of the hydrogen atom when only nucleus I exists and $\psi_{II}(\mathbf{x})$ is the wave function of the hydrogen atom when only nucleus II exists. We assume that both hydrogen atoms are in the ground state.

In practice, however, nucleus I or nucleus II does not exist alone, but the two nuclei exist with a finite separation. Even then, if their separation is very large, we can consider either $\psi_I(\mathbf{x})$ or $\psi_{II}(\mathbf{x})$ as the eigenfunction of the whole system in a zeroth approximation. We can also assume that they belong to the same eigenvalue in the zeroth approximation. Therefore, the state of this system is doubly degenerate in this approximation, and we can use a linear combination of $\psi_I(\mathbf{x})$ and $\psi_{II}(\mathbf{x})$

$$\phi(\mathbf{x}) = C_1 \psi_I(\mathbf{x}) + C_2 \psi_{II}(\mathbf{x}) \tag{10-1}$$

as the eigenfunction for the whole system in the zeroth approximation.

However, even if the internuclear distance is large and $\psi_I(\mathbf{x})$ and $\psi_{II}(\mathbf{x})$ are still good approximations for the eigenfunctions, if the distance is finite, the double degeneracy of the eigenvalue is broken, and the level should split into two. Then for each eigenvalue, C_1 and C_2 should be determined uniquely (apart from a constant factor). In order to determine the values of C_1 and C_2, we can use perturbation theory for the degenerate levels, but it is more convenient for our purpose to use a different method.

Please put up with my bad habit of writing a few formulas. Let us assume that the two nuclei are at points $-\mathbf{r}/2$ and $\mathbf{r}/2$ on the x-axis. (Needless to say, $\pm \mathbf{r}/2$ are the vectors with the components $(\pm r/2, 0, 0)$, and the internuclear distance is r.) Then the Schrödinger equation of our system is

$$\left[\frac{1}{2m} \mathbf{p}^2 - \frac{e^2}{|\mathbf{x} + \frac{\mathbf{r}}{2}|} - \frac{e^2}{|\mathbf{x} - \frac{\mathbf{r}}{2}|} - E \right] \phi(\mathbf{x}) = 0. \tag{10-2}$$

If we make the transformation $\mathbf{x} \rightarrow -\mathbf{x}$, we obtain the equation for $\phi(-\mathbf{x})$, but the expression inside the brackets of (10-2) is invariant. Therefore, we are led to the conclusion that if $\phi(\mathbf{x})$ is a solution, then $\phi(-\mathbf{x})$ is also a solution. Please remember lecture 5, in which we derived $(5\text{-}16)_s$ and $(5\text{-}16)_a$ from (5-14). Then we find that for the eigenfunction of (10-2), either

$$\phi(\mathbf{x}) = \phi(-\mathbf{x}) \equiv \phi_s(\mathbf{x}) \quad \text{or} \tag{10-3$_s$}$$

$$\phi(\mathbf{x}) = -\phi(-\mathbf{x}) \equiv \phi_a(\mathbf{x}). \tag{10-3$_a$}$$

In other words, the eigenfunction of (10-2) is a function either symmetric or antisymmetric with respect to the transformation $\mathbf{x} \leftrightarrow -\mathbf{x}$.

The relations $(10\text{-}3)_s$ and $(10\text{-}3)_a$ thus obtained are valid for large as well as small internuclear distance r. Let us limit ourselves to the case in which r is sufficiently large that $\phi(\mathbf{x})$ in the form of (10-1) is still a good approximation.

For such a case the relations (10-3)$_a$ and (10-3)$_s$ are valid only when we use $C_1 = C_2$ and $C_1 = -C_2$, respectively, in the right-hand side of (10-1). Namely, if we consider the wave function for the ground state of a hydrogen atom placed at the origin, then it is a function of the form $\psi(|\mathbf{x}|)$, and if we use this function to express $\psi_I(\mathbf{x})$ and $\psi_{II}(\mathbf{x})$, then

$$\psi_I(\mathbf{x}) = \psi\left(\left|\mathbf{x} + \frac{\mathbf{r}}{2}\right|\right) \tag{10-4}$$

$$\psi_{II}(\mathbf{x}) = \psi\left(\left|\mathbf{x} - \frac{\mathbf{r}}{2}\right|\right),$$

and therefore we see that the transformation $\mathbf{x} \leftrightarrow -\mathbf{x}$ is equivalent to the transformation I $\leftrightarrow$ II, and finally

$$\phi_s(\mathbf{x}) = \frac{1}{\sqrt{2}}[\psi_I(\mathbf{x}) + \psi_{II}(\mathbf{x})] \tag{10-4$_s$}$$

$$\phi_a(\mathbf{x}) = \frac{1}{\sqrt{2}}[\psi_I(\mathbf{x}) - \psi_{II}(\mathbf{x})] \tag{10-4$_a$}$$

satisfy (10-3)$_s$ and (10-3)$_a$, respectively.

Now, how about the eigenvalue of (10-2)? First notice that since r appears as a parameter in (10-2), the eigenvalue $E(r)$ is also a function of r. As I told you earlier, if the distance between the nuclei is finite, the eigenvalues split into two. Therefore, $\phi_s(r)$ and $\phi_a(r)$ have the eigenvalues $E_s(r)$ and $E_a(r)$, respectively. When the Coulomb energy e^2/r between the protons is added to these two eigenvalues, the resulting quantities

$$J_s(r) = E_s(r) + \frac{e^2}{r} \tag{10-5}$$

$$J_a(r) = E_a(r) + \frac{e^2}{r}$$

give the total energies for the $\phi_s(\mathbf{x})$ and $\phi_a(\mathbf{x})$ states, respectively, and therefore they each become a potential for the force acting between nuclei I and II. We can show in general, regardless of the value of r, that $J_s(r) < J_a(r)$, and furthermore for large values of r for which (10-4)$_s$ and (10-4)$_a$ are good approximations, $J_s(r)$ is negative, and $-J_a(r) = J_s(r)$. We can therefore write

$$J_s(r) = -J(r) \tag{10-5'}$$

$$J_a(r) = J(r),$$

where $J(r) > 0$, monotonically decreasing toward $J(\infty) = 0$ as r is increasing. We therefore see that there is an attraction between nucleus I and nucleus II in the symmetric state and a repulsion between them in the antisymmetric state. Thus H_2^+ will be stable in the symmetric state but not in the antisymmetric state.

We thus see that two protons are stably bound in spite of the Coulomb repulsion because of the electron surrounding the nuclei. As our eigenfunction $(10\text{-}4)_s$ or $(10\text{-}4)_a$ indicates, $\psi_I(\mathbf{x})$ or $\psi_{II}(\mathbf{x})$ alone does not give a correct solution even if the two nuclei are far apart. The electron is visiting both nuclei equally, and if the way of visiting is symmetric, there will be an attraction between the two nuclei, and if it is antisymmetric, there will be repulsion. This will be our conclusion.

We naively thought at the outset that, if the two nuclei are infinitely far apart, $\psi_I(\mathbf{x})$ and $\psi_{II}(\mathbf{x})$ can each individually be an eigenfunction; but this seems to contradict the fact that the electron is visiting both nuclei equally. How can we understand this? You will be relieved of this worry if you think as follows.

Let us construct a wave packet

$$\phi_\pm(\mathbf{x}, t) = \phi_s(\mathbf{x})e^{-iJ_s(r)t/\hbar} \pm \phi_a(\mathbf{x})e^{-iJ_a(r)t/\hbar} \qquad (10\text{-}6)$$

using the eigenfunctions $\phi_s(\mathbf{x})$ and $\phi_a(\mathbf{x})$ given in $(10\text{-}4)_s$ and $(10\text{-}4)_a$. This is certainly a solution of the time-dependent Schrödinger equation, and therefore this is a possible state of our mechanical system. Let us substitute $(10\text{-}4)_s$ and $(10\text{-}4)_a$ for $\phi_s(\mathbf{x})$ and $\phi_a(\mathbf{x})$, respectively, on the right-hand side of (10-6) and take the absolute square of both sides. We immediately obtain

$$|\phi_\pm(\mathbf{x}, t)|^2 = |\psi_I(\mathbf{x})|^2 \left[1 \pm \cos\frac{1}{\hbar}(J_a - J_s)t\right] +$$
$$|\psi_{II}(\mathbf{x})|^2 \left[1 \mp \cos\frac{1}{\hbar}(J_a - J_s)t\right] + \dots \qquad (10\text{-}6')$$

Here the ellipsis includes terms in $\psi_I^*(\mathbf{x})\psi_{II}(\mathbf{x})$ and $\psi_I(\mathbf{x})\psi_{II}^*(\mathbf{x})$, which go to zero as $r \to \infty$, and we can ignore them if r is sufficiently large. If these terms are neglected, the right-hand side of (10-6') shows that the coefficients of $|\psi_I(\mathbf{x})|^2$ and $|\psi_{II}(\mathbf{x})|^2$ increase and decrease alternately with the frequency

$$\nu \equiv \frac{1}{h}[J_a(r) - J_s(r)] = \frac{2}{h}J(r). \qquad (10\text{-}7)$$

This means that the electron jumps back and forth from I to II and II to I with the frequency ν. From the relation $J(\infty) = 0$ just mentioned, this frequency becomes zero as $r \to \infty$. Therefore, in this limit ϕ_+ is always ψ_I, and ϕ_- is always ψ_{II}. We now see that the contradiction mentioned earlier has been resolved. At the same time, we now understand that there is the intimate relation (10-7) between the potential $\pm J(r)$ of the force acting between the two nuclei and the frequency ν of the electron shuttling between the two nuclei. If the electron did not move from one nucleus to the other, then the frequency would be zero, and therefore $J(r)$ would also be zero.

We therefore can say the exchange force originates in the shuttling of the electron; if we express this shuttling of the electron in other words, we may also imagine that nucleus I and nucleus II alternately don the electron. Or we can say that the particle at $-\mathbf{r}/2$ and the particle at $\mathbf{r}/2$ are alternately becoming the neutral atom and the positive ion. This *alternation* is the essence of the exchange force.

Heisenberg thought something similar is taking place between the neutrons and protons in nuclei. Namely, a proton and a neutron in the nucleus also alternately gain or lose the charge, and thus a neutron changes to a proton and a proton changes to a neutron repeatedly. This was Heisenberg's idea. He thought that if this exchange of electric charge occurs with a frequency ν, then there should appear a potential $\pm J(r)$ given by (10-7).

Here we obviously cannot adopt the analogy with H_2^+ literally. This is because in the case of H_2^+ the exchange of an electron not only changes a hydrogen atom into a hydrogen ion and a hydrogen ion into a hydrogen atom, but, since the Ehrenfest-Oppenheimer rule shows that the hydrogen atom is a boson and the hydrogen ion is a fermion, the statistics is also alternating from Bose to Fermi and Fermi to Bose. However, in the case of a neutron and a proton, both are fermions, and the statistics is constant. But if we can consider the neutron to be composed of a proton and a particle with spin zero which may be called a boson electron (this is a version of Rutherford's neutron), then it is not impossible to imagine that this boson electron is going back and forth between the neutron and the proton. Heisenberg did refer to this idea. Nevertheless, he concluded that it was better to ignore the existence of the boson electron, perhaps because he was not sure whether he could use quantum mechanics for the shuttling of this particle even if this idea were adopted; he did not adopt this idea and deemed the neutron and proton as equally elementary particles, for which the alternation of neutron to proton and proton to neutron with a frequency of $2J(r)/h$ was an intrinsic property of these particles.

How can we mathematically formulate the exchange force without introducing a gimmick like the boson electron? In the case of H_2^+, the symmetry of the wave function with respect to the transformation $\mathbf{x} \rightarrow -\mathbf{x}$ determined whether the potential of the exchange force was $-J(r)$ or $+J(r)$. In this case, $\mathbf{x}$ was the coordinate of the electron (or boson electron). Therefore, we cannot use this formulation without considering the electron (boson electron) for the exchange force. What can we do? Here Heisenberg was led to the concept of *isospin* from the analogy with spin and established a formulation of the exchange force without using $\mathbf{x}$.

Remember that in lecture 5 I discussed the apparently strong interaction between two spins. When just now I talked about H_2^+, I reminded you about

(5-14), (5-16)$_s$, and (5-16)$_a$; as in the case of H$_2^+$, the symmetric and antisymmetric states also appeared there, and the energy difference resulted from the symmetry of the state, and this difference was attributed to the apparent spin-spin interaction. I did not use the term *exchange force* in lecture 5, but this apparent force is *precisely* the exchange force. Although this exchange force originates from the orbital angular momentum of the electrons, we can regard it as if it were an interaction between two spins.

As I told you in lecture 5, it was Heisenberg who played the major role in the discussion of the spin-spin interaction. Therefore, it is understandable that he thought that the exchange force in the nucleus can be formulated using something like the spin matrix regardless of whether or not the boson electron exists. Now we shall go into the story of isospin.

When I discussed the problem of H$_2^+$, I said that the particle at the position $-\mathbf{r}/2$ and the particle at position $\mathbf{r}/2$ alternately become the neutral atom and the positive ion. Now if the particle at $-\mathbf{r}/2$ is the neutral atom and that at $\mathbf{r}/2$ is the positive ion, then the state is ψ_I; if the particle at $-\mathbf{r}/2$ is the positive ion and the particle at $\mathbf{r}/2$ is the neutral atom, then the state is ψ_{II}. In in (10-4)$_s$ ϕ_s has the property that it does not change sign when the neutral atom is changed to the positive ion and the positive ion is changed to the neutral atom, and ϕ_a of (10-4)$_a$ has the property of changing sign. We can say this: "ϕ_s is symmetric and ϕ_a is antisymmetric with respect to the transformation which changes the neutral atom to the positive ion and the positive ion to the neutral atom." We have learned that there exist the attraction $-J(r)$ for ϕ_s and the repulsion $+J(r)$ for ϕ_a. This last point discriminates between the exchange force and an ordinary force, i.e., for an ordinary force the same potential appears regardless of the symmetry of ϕ.

So far I have been discussing H$_2^+$, but if the exchange force is defined in this way, then the coordinate $\mathbf{x}$ of the electron is not explicitly involved, and therefore we can consider this to be the characteristic of the exchange force between a neutron and a proton. In order to do that, we simply replace "neutral atom" for H$_2^+$ with "neutron" and "positive ion" with "proton." This is Heisenberg's idea.

Now then, how can we mathematically express the operation of changing a proton into a neutron and a neutron into a proton? At this very point Heisenberg introduced isospin. For this purpose, instead of considering the neutron and proton as different elementary particles, he considered them as two different states of the same elementary particle. This elementary particle was later called *nucleon*. If I use this terminology, I can say that the constituents of an atomic nucleus are elementary particles called nucleons which have two states called the neutron state and the proton state. Since the neutron is a fermion with spin 1/2, in both states the nucleon is a fermion with spin 1/2, and therefore we can regard the nucleon itself to be a fermion with spin 1/2. If we maintain that the nucleon can take two states, then automatically we must accept a fifth degree of freedom in addition to the x, y, z, and spin degrees of freedom.

This fifth degree of freedom is like the spin degree of freedom in that it can take only two states. Therefore, Heisenberg introduced the 2×2 matrices

$$\rho^\xi = \begin{pmatrix} 0 & 1 \\ 1 & 0 \end{pmatrix} \quad \rho^\eta = \begin{pmatrix} 0 & -i \\ i & 0 \end{pmatrix} \quad \rho^\zeta = \begin{pmatrix} 1 & 0 \\ 0 & -1 \end{pmatrix} \qquad \text{(10-8)}$$

to describe this degree of freedom. Obviously these have the identical form as the Pauli spin matrices. From the analogy with the case of spin in which spin up is $\sigma_z = +1$ and spin down is $\sigma_z = -1$, Heisenberg chose the state for which the eigenvalue of ρ^ζ is $+1$ to be the neutron state and the state for which the eigenvalue is -1 to be the proton state.[2] Then if we write the wave function (which is a two-component quantity) for ρ as

$$\alpha = \begin{pmatrix} \alpha_1 \\ \alpha_2 \end{pmatrix}, \qquad \text{(10-9)}$$

then

$$\alpha^n \equiv \begin{pmatrix} 1 \\ 0 \end{pmatrix} \quad \text{and} \quad \alpha^p \equiv \begin{pmatrix} 0 \\ 1 \end{pmatrix} \qquad \text{(10-9')}$$

represent the eigenfunctions of the neutron state and the proton state, respectively. Let me note that just as in the case of spin, there are relations

$$\begin{aligned} \rho^\xi \alpha^n &= \alpha^p & \rho^\xi \alpha^p &= \alpha^n \\ \rho^\eta \alpha^n &= i\alpha^p & \rho^\eta \alpha^p &= -i\alpha^n \\ \rho^\zeta \alpha^n &= \alpha^n & \rho^\zeta \alpha^p &= -\alpha^p. \end{aligned} \qquad \text{(10-10)}$$

Now let us consider two nucleons and distinguish them by the subscripts I and II. Then the wave function for the state in which nucleon I is a neutron and nucleon II is a proton can be expressed as the product

$$\alpha^{n,p} = \alpha_I^n \alpha_{II}^p, \qquad \text{(10-11)}_{n,p}$$

and that for the state in which I is a proton and II is a neutron can be expressed as

$$\alpha^{p,n} = \alpha_I^p \alpha_{II}^n, \qquad \text{(10-11)}_{p,n}$$

If I continue the analogy of H_2^+, $\alpha^{n,p}$ corresponds to ψ_I, and $\alpha^{p,n}$ corresponds to ψ_{II}. We now see the transformation "changing a neutron into a proton and

2. The modern convention is the opposite of Heisenberg's, i.e., $+1$ for the proton state and -1 for the neutron state.

a proton into a neutron" is nothing but changing n to p and p to n in these formulas. Next we can write the formulas corresponding to ϕ_s and ϕ_a for H_2^+. They are

$$\alpha^s = \alpha_I^n \alpha_{II}^p + \alpha_I^p \alpha_{II}^n \qquad (10\text{-}12)_s$$

$$\alpha^a = \alpha_I^n \alpha_{II}^p - \alpha_I^p \alpha_{II}^n \qquad (10\text{-}12)_a$$

Here it is obvious that α^s is symmetric with respect to the transformation that changes a neutron into a proton and a proton into a neutron, and α^a is antisymmetric.

Having come this far, we can immediately make a matrix which is $-J(r)$ for a symmetric state and $+J(r)$ for an antisymmetric state. Namely,

$$-\frac{1}{2}\left(\rho_I^\xi \rho_{II}^\xi + \rho_I^\eta \rho_{II}^\eta\right) J(r) \qquad (10\text{-}13)$$

has this property. If what I just mentioned is the only requirement, then the expression in parentheses could be either $\rho_I^\xi \rho_{II}^\xi$ only or $\rho_I^\eta \rho_{II}^\eta$ only. But Heisenberg thought there was no nuclear force between protons and between neutrons, so he had to resort to (10-13) so that, if both nucleons were in the neutron state or if both were in the proton state, there would be no nuclear force. Using (10-10), we can easily see that

$$\frac{1}{2}\left(\rho_I^\xi \rho_{II}^\xi + \rho_I^\eta \rho_{II}^\eta\right)\alpha^s = \alpha^s$$

$$\frac{1}{2}\left(\rho_I^\xi \rho_{II}^\xi + \rho_I^\eta \rho_{II}^\eta\right)\alpha^a = -\alpha^a$$

$$\frac{1}{2}\left(\rho_I^\xi \rho_{II}^\xi + \rho_I^\eta \rho_{II}^\eta\right)\alpha_I^n \alpha_{II}^n = 0 \qquad (10\text{-}14)$$

$$\frac{1}{2}\left(\rho_I^\xi \rho_{II}^\xi + \rho_I^\eta \rho_{II}^\eta\right)\alpha_I^p \alpha_{II}^p = 0.$$

From the first two equations of (10-14), we find that in the α^s state, (10-13) has the value of $-J(r)$, and in the α^a state it has $+J(r)$; and from the last two equations of (10-14), we can conclude that there is no force if the nucleons are both neutrons or both protons. Furthermore, by using the "wave packet"

$$\alpha^s e^{-iJ_s(r)t/\hbar} \pm \alpha^a e^{-iJ_a(r)t/\hbar},$$

we find that the transformation from a neutron into a proton or a proton into a neutron occurs with the frequency $\nu = 2J(r)h$.

So far we have considered only two nucleons, but if there exist, in general,

N nucleons, then we can consider a matrix

$$-\frac{1}{2} \sum_{K>L}^{N} \left(\rho_K^\xi \rho_L^\xi + \rho_K^\eta \rho_L^\eta \right) J(r_{KL}) \tag{10-15}$$

which expresses the sum of the exchange forces. In addition to this exchange force, there is a Coulomb force between protons, so we simply have to add

$$+\frac{1}{4} \sum_{K>L}^{N} \left(1 - \rho_K^\zeta \right) \left(1 - \rho_L^\zeta \right) \frac{e^2}{r_{KL}}. \tag{10-15'}$$

Therefore, as the total Hamiltonian for a nucleus, we can use

$$H = \frac{1}{2m} \sum_K^N \mathbf{p}_K^2 - \frac{1}{2} \sum_{K>L}^{N} \left(\rho_K^\xi \rho_L^\xi + \rho_K^\eta \rho_L^\eta \right) J(r_{KL}) +$$
$$\frac{1}{4} \sum_{K>L}^{N} \left(1 - \rho_K^\zeta \right) \left(1 - \rho_L^\zeta \right) \frac{e^2}{r_{KL}}. \tag{10-16}$$

This is Heisenberg's idea. Since the mass of a nucleon is large, the velocity of a nucleon in the nucleus is small, and a nonrelativistic Hamiltonian suffices.

From the Hamiltonian (10-16), Heisenberg derived a variety of conclusions. Furthermore, since the simplest deuteron presents only a two-body problem, it is possible to provide a mathematically rigorous treatment. Doing so, we find a few points to be corrected in the Hamiltonian (10-16). One of them is that we must change the sign of $J(r)$ in (10-16). For as it is, the spin of the deuteron becomes zero. (When he wrote this paper, Murphy and Johnston had not yet done their experiment.) The other point is that if we use his Hamiltonian, the great stability of the helium nucleus cannot be explained, and the repulsive force (because we changed the sign of $J(r)$) in the ϕ_s state of the deuteron becomes too big and does not agree with the collisional experiments between the neutron and the proton. This last point, however, can be improved by adding a term $(\sigma_K \cdot \sigma_L)$, which contains spin, to the exchange force. [This nuclear force containing $(\sigma_K \cdot \sigma_L)$ is named the *Majorana force* after its proposer.]

Around 1936 the remarkable fact was shown experimentally that even between two protons there is a nuclear force in addition to the Coulomb force, and this force equals the force between a neutron and a proton in the ϕ_s state. If we thus assume that the same force also acts between neutrons, then this remarkable fact can be incorporated into the theory by substituting $(\rho_K^\xi \rho_L^\xi + \rho_K^\eta \rho_L^\eta + \rho_K^\zeta \rho_L^\zeta) = (\rho_K \cdot \rho_L)$ for the factor $(\rho_K^\xi \rho_L^\xi + \rho_K^\eta \rho_L^\eta)$ in (10-15). Here I wrote ρ, which means we considered $(\rho^\xi, \rho^\eta, \rho^\zeta)$ as a vector in

Figure 10.2 Ettore Majorana (1906–1938), 1923 university document-card photograph. [Courtesy of AIP Emilio Segrè Visual Archives]

$\xi\eta\zeta$ space, but the fact that $\boldsymbol{\rho}^\xi$, $\boldsymbol{\rho}^\eta$, $\boldsymbol{\rho}^\zeta$ appear in the expression for the nuclear force in the form of the inner product $(\boldsymbol{\rho}_K \cdot \boldsymbol{\rho}_L)$ means that as far as the nuclear force is concerned, $\xi\eta\zeta$ space is isotropic just as xyz space, in which the spin components σ^x, σ^y, σ^z are defined, is isotropic. From these points of view, $\boldsymbol{\rho}$ is not just a convenient tool to formulate the nuclear force, but like σ it is a very fundamental physical quantity. For this reason, just as we called $\sigma/2$ *spin*, we call $\boldsymbol{\rho}/2$ *isospin*. If we adopt this idea, the occurrence of the two kinds of potential, $J_s(r)$ or $J_a(r)$, depends on whether the isospins of nucleons I and II are parallel or antiparallel.

Now let me give you just one example in which the concept of isospin plays a role. If we extend Heisenberg's idea, namely that the neutron and the proton are not different elementary particles but different states of the same elementary particle, i.e., the nucleon, then the three nuclear species ^{14}C, ^{14}N, and ^{14}O may be regarded as different states of the same nucleus, since they are all composed of fourteen nucleons. In the approximation in which the Coulomb interaction is neglected, the total isospin and its ζ-component commute with the Hamiltonian, and therefore

$$\left(\sum_{K=1}^{N} \frac{1}{2}\boldsymbol{\rho}_K\right)^2 = T(T+1) \quad \text{and} \quad \sum_{K=1}^{N} \frac{1}{2}\rho_K^\zeta = T^\zeta$$

are two conserved quantities, and we can use the numbers T and T^ζ related to their eigenvalues as the quantum numbers to label the state of the nuclei. This is analogous to the case of atomic spectroscopy in which s of $(\sum_K \frac{1}{2}\sigma_K)^2 = s(s+1)$ and s^z from $\sum_K \frac{1}{2}s_K^z = s^z$ are used to label atomic levels.

If we apply this idea to the case of ^{14}C, ^{14}N, ^{14}O, then from a variety of experiments we see that the ground state of ^{14}N is the state for $T = 0$, $T^\zeta = 0$ and that the ground state of ^{14}C, the first excited state of ^{14}N, and the ground state of ^{14}O correspond to the three states of $T = 1$. Out of these three states, the state of ^{14}C corresponds to $T^\zeta = -1$, that of ^{14}N to $T^\zeta = 0$, and that of ^{14}O to $T^\zeta = 1$. Therefore, if we use spectroscopic terminology, the ground state of ^{14}N is singlet and the remaining three states are triplet. The three levels of the triplet are degenerate if we do not consider the Coulomb force between protons; but if we do consider it, then the triplet splits into three levels, just as in atomic spectroscopy the levels are split into three levels by the internal magnetic field. Furthermore, the splitting pattern agrees very well between experiment and calculation. I show this pattern in figure 10.3. As you see from this example, it has become possible to relate energy levels of nuclei which are isobaric to each other by applying to isospin the same method that was used for treating multiplets in atomic spectroscopy. In this sense, isospin is sometimes also called isobaric spin. (Somebody[3] called it isotopic spin, but I do not think this is a good term.)

The concept of isospin plays an important role not only in this nuclear "spectroscopy" but also in the Fermi theory and the Yukawa theory, as I shall

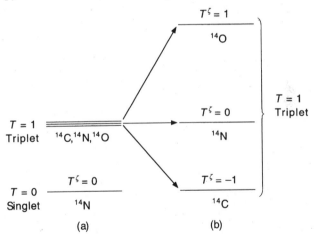

Figure 10.3 Energy levels of the ^{14}C, ^{14}N, and ^{14}O nuclei. (a) shows the levels when the Coulomb splitting is neglected; (b) shows the splitting of the triplet state when the Coulomb energy is taken into account.

3. Eugene Wigner.

explain later. The fact that this concept has come to have an initially unexpected meaning clearly shows how prescient Heisenberg was. Heisenberg arrived at the idea of isospin from the analogy with spin, and I think one of the major characteristics of his thought process is this type of analogy. Moreover, his analogizing does not remain merely a phenomenological method, but in many cases it turns out to hit upon something fundamental. Dirac's acrobatics, Pauli's frontal assault, and Heisenberg's analogizing: each is uniquely characteristic of its practitioner so that we are never bored following their work.

Heisenberg's theory opened the way for the application of quantum mechanics to problems inside the nucleus. As I told you in the previous lecture, the statistics, spin, and magnetic moment of the neutron were obtained from those of the deuteron, and the sole reason that this was possible was that quantum mechanics could be applied to the deuteron. Furthermore, for the statistics of nuclei, we can use the Ehrenfest-Oppenheimer rule to conclude that if A is even (odd), then the nucleus follows Bose (Fermi) statistics; on the other hand for spin, according to the addition rule for angular momenta, if there are an even (odd) number of spin $1/2$ particles, their total angular momentum is integral (half-integral). If we summarize these two conclusions, then we have that if the spin is integral (half-integral), then the nucleus follows Bose (Fermi) statistics. This means that Pauli's relation, which we discussed in lecture 8, also holds true for composite particles. Moreover, we can use isospin instead of spin in these rules. In our example of ^{14}C, ^{14}N, ^{14}O, there are singlet and triplet states, and this means that the total isospin is an integer and follows the rule. These conclusions could be reached because quantum mechanics *is* applicable inside the nucleus.

However, even Heisenberg apparently was not free from Bohr's influence, and he dared not touch the problem of β-decay, believing that quantum mechanics was not applicable. It was none other than Fermi who rejected Bohr's idea and incorporated β-decay into the framework of quantum mechanics utilizing Pauli's neutrino. His work appeared in 1934.

Fermi started from Heisenberg's idea that there is no such thing as an electron in the nucleus and that there exist only several nucleons which have the fifth degree of freedom of $\rho^\zeta = \pm 1$. Let us consider a nucleus with mass number A. This is a mechanical system composed of A nucleons. There exist a variety of nuclei with different atomic numbers as isobars, but let us take the Z of the lowest energy as Z_0. Then among the isobars, those with $Z \geq Z_0 + 1$ or $Z \leq Z_0 - 1$ can be regarded as excited states of our nucleon system. (In the example of a fourteen-nucleon system which I just mentioned, the ground states of ^{14}C and ^{14}O were regarded as excited states of ^{14}N.) Now if we consider a nucleus with $Z = Z_0 - 1$, then this nucleus has one more nucleon in the neutron state and one less in the proton state than the Z_0 nucleus. If the excitation energy is larger than the rest mass of the electron, mc^2, then the nucleus undergoes a transition

Figure 10.4 Fermi (1901–1954) and Bohr, 1931. [Photograph by S. A. Goudsmit. Courtesy of AIP Emilio Segrè Visual Archives]

to the ground state Z_0. In the process one of the nucleons in the neutron state emits an electron and Pauli's neutrino. (We assume here the neutrino mass to be 0.) This must occur. Fermi thought this was β-decay.

How did he incorporate this idea into quantum mechanics? In order to answer this question, one can use the example of the well-known process in which an excited atom relaxes to the ground state by emitting a photon. In order to treat this problem, we first quantize the radiation field as we did in lecture 6, include in the Hamiltonian the interaction between the quantized field and the electrons in the atom, set up a Schrödinger equation, and solve it. Indeed, what Fermi did was an extension of this procedure.

First, Fermi quantized the electron field and the neutrino field. He already knew the quantization of the electron field, but he had to determine whether the neutrino field was a boson or a fermion. However, since both the neutron and the proton have spin 1/2, the angular momentum changes by integral values in the transition from the neutron state to the proton state. From this and from the conservation law of angular momentum, it is inferred that the value of the total angular momentum of the emitted electron and neutrino also must be an integer, and therefore in order to obtain an integral value when combining with the electron angular momentum, the spin of the neutrino is required to be half-integral. Fermi assumed that the spin of the neutrino is 1/2 because it is the simplest choice. Then the neutrino satisfies the Dirac equation, and in order to avoid the difficulty of negative energy, it must be a fermion.

After finding the equation and statistics satisfied by the neutrino, the quantization of the field is straightforward. Also, the form of the interaction between the electron-neutrino field and the nucleon is restricted by the requirements of covariance and of charge conservation. The interaction energy should contain the components of isospin $\rho^\xi = \begin{pmatrix} 0 & 1 \\ 1 & 0 \end{pmatrix}$ and $\rho^\eta = \begin{pmatrix} 0 & -i \\ i & 0 \end{pmatrix}$. Otherwise, the matrix element of the interaction coupling the neutron state and the proton state would be zero, and the transition neutron state ↔ proton state would not occur. I shall just show you what Fermi wrote down as the simplest interaction energy satisfying these conditions:

$$g \sum_K V(\mathbf{x}_K) = g \sum_K \left[\frac{1}{2} \left(\rho^\xi_K - i\rho^\eta_K \right) \Psi^\dagger(\mathbf{x}_K) \Phi(\mathbf{x}_K) + \frac{1}{2} \left(\rho^\xi_K + i\rho^\eta_K \right) \Phi^\dagger(\mathbf{x}_K) \Psi(\mathbf{x}_K) \right]. \quad (10\text{-}17)$$

Here $\Psi(\mathbf{x})$ is the wave function of the quantized electron, $\Phi(\mathbf{x})$ is the wave function of the quantized antineutrino, and both of them have four components that satisfy the Dirac equation. The notation $\Psi^\dagger(\mathbf{x}_K)\Phi(\mathbf{x}_K)$ or $\Phi^\dagger(\mathbf{x}_K)\Psi(\mathbf{x}_K)$ means $\sum_{\alpha=1}^4 \Psi^\dagger_\alpha(\mathbf{x}_K)\Phi_\alpha(\mathbf{x}_K)$ or $\sum_{\alpha=1}^4 \Phi^\dagger_\alpha(\mathbf{x}_K)\Psi_\alpha(\mathbf{x}_K)$; $\mathbf{x}_K$, ρ^ξ_K, ρ^η_K, needless to say, are the spatial coordinates and isospin of nucleon K; and g is the coupling constant.

In the back of Fermi's mind when he chose (10-17) as the interaction energy was the analogy with the radiation field. In the case of a charged particle and the radiation field, the interaction is $e \sum_K V(\mathbf{x}_K)$, where $V(\mathbf{x}_K)$ is the potential of the electric field describing the radiation field as long as the velocity of the particle is smaller than the velocity of light c. Fermi's equation (10-17) generalizes this to make it applicable to the case of isobaric nuclei. Namely, in (10-17) the products $\Psi^\dagger(\mathbf{x}_K)\Phi(\mathbf{x}_K)$ or $\Phi^\dagger(\mathbf{x}_K)\Psi(\mathbf{x}_K)$ are used instead of

$V(\mathbf{x}_K)$ for the case of the charged particle and the radiation field. This is because only one particle, the photon, is associated with the radiation field, whereas two particles, the electron and the neutrino, are associated with the field for β-decay. Furthermore, the expressions $(\rho_K^\xi - i\rho_K^\eta)/2$ and $(\rho_K^\xi + i\rho_K^\eta)/2$ in (10-17) correspond to the transitions neutron $\leftrightarrow$ proton. Indeed, from (10-8) we find $(\rho_K^\xi - i\rho_K^\eta)/2 = \begin{pmatrix} 0 & 0 \\ 1 & 0 \end{pmatrix}_K$ and $(\rho_K^\xi + i\rho_K^\eta)/2 = \begin{pmatrix} 0 & 1 \\ 0 & 0 \end{pmatrix}_K$, which correspond to the matrices for the transitions neutron $\rightarrow$ proton and proton $\rightarrow$ neutron, respectively.

We considered the case where $Z = Z_0 - 1$, but for $Z = Z_0 + 1$, β-decay is also possible if energy allows it. The difference here though is that a positron $+e$ and a neutrino ν are emitted. Indeed, for the case of ^{14}C, ^{14}N, and ^{14}O discussed earlier, two types of β-decay are observed: ^{14}C $\rightarrow$ ^{14}N and ^{14}O $\rightarrow$ ^{14}N. The former is accompanied by electron emission and the latter by positron emission.

Thus Fermi succeeded in introducing β-decay into the framework of quantum mechanics. Pauli's concept of the neutrino was too far ahead of its time, but triggered by Chadwick's discovery and Heisenberg's theory, it was taken up by Fermi four years later. Therefore what Pauli called a "desperate remedy" was in fact a promising one.

Stimulated by the success of Fermi's theory, there arose the idea that Fermi's method might be applicable to the problem which Heisenberg avoided, namely whether the exchange of some particle was responsible for the exchange force; to realize this picture, the exchange of an electron-neutrino pair might be used instead of the boson electron. However, these attempts all failed because no one could account for the strong nuclear force observed in experiments. To elaborate, the strength of the exchange force is proportional to the frequency of the shuttling of the particles, and therefore to get a strong exchange force we need an extremely frequent shuttling of the electron-neutrino pair. Then the probability of the electron-neutrino pair escaping from the nucleus was proportionately large, and therefore the lifetime of the nucleus undergoing β-decay would become orders of magnitude shorter than is observed by experiments.

Therefore, Yukawa arrived at the idea that there should exist a yet-to-be-discovered charged boson, the *heavy quantum*, and that this charged particle shuttles between the neutron and proton. He concluded that if this particle has a mass about 100 times as large as that of the electron, then the effective range of the nuclear force is on the order of 10^{-13} cm. From the analogy that the force between charged particles is mediated by the electromagnetic field, he thought that the nuclear force is mediated by an unknown field which might be called the nuclear force field. When he adopted the Klein-Gordon equation as the equation of this field, then instead of the Coulomb potential e^2/r for the electromagnetic field, $g^2 e^{-\kappa r}/r$ appeared, and from the de Broglie–Einstein

Figure 10.5 Yukawa (1907–1981).
[Courtesy of AIP Emilio Segrè Visual
Archives, E. Scott Barr Collection]

relation, in place of the zero-mass boson (photon), which is the quantized electromagnetic field, he obtained a boson of mass $\kappa h/c$ for the nuclear force field. As e^2/r is the potential for the Coulomb force, $g^2 e^{-\kappa r}/r$ will be the potential for the nuclear force, he thought, and calculated the boson mass by putting 10^{-13} cm for the force range $1/\kappa$, obtaining 100 times the electron mass that I mentioned above. Thus this boson is lighter than the proton but heavier than the electron and much heavier than the photon. He therefore named this particle the *heavy quantum* as opposed to the *light quantum*. Even Yukawa, it seems, played with puns.

Further, Yukawa thought that this nuclear-force boson would decay into an electron (or a positron) and a neutrino with a lifetime on the order of 10^{-6} s, and he explained the β-decay of nuclei as follows. In Fermi's theory the process neutron ↔ proton generates the electron-neutrino pair, and therefore we have the difficulty of a short nuclear lifetime when this frequency increases. But in Yukawa's theory, the process of neutron ↔ proton is effected by the shuttling of the nuclear-force boson, and β-decay occurs during this process when the boson—on a lark—breaks down into an electron and a neutrino. If this is rare, then β-decay seldom happens, and therefore we need not worry about a brief nuclear lifetime.

You may worry about the next point. Namely, if the nuclear-force boson shuttles between the nucleons with such high frequency necessary to explain the large nuclear force, then the probability that the boson itself escapes from the nucleus also becomes large, and the nucleus may be observed to disintegrate, emitting this boson. The answer to this is quite simple. This boson has a mass 100 times as large as that of the electron so that its mc^2 is 10^8 eV. Therefore, unless this large energy is supplied, it cannot come out of the nucleus. Such a highly excited nucleus does not naturally form, and particle accelerators of such high energy were not known at that time.

However, among the cosmic rays, there are many particles with kinetic energy on the order of 10^8 eV. If such a particle collides with an atomic nucleus, then possibly the nuclear-force boson will be knocked out of that nucleus. Therefore, Yukawa suggested that this particle may be discovered among cosmic rays. In 1936, a year after Yukawa's paper appeared, Anderson and Neddermeyer discovered among cosmic rays a particle which resembled that described by Yukawa. This particle was named *meson* (intermediate particle).[4]

By the way, I might mention here that isospin plays an important role in Yukawa's theory also. Yukawa used for the interaction energy of nucleons and the nuclear force field

$$g \sum_K \left[\frac{1}{2} \left(\rho_K^\xi - i \rho_K^\eta \right) U^\dagger(\mathbf{x}_K) + \frac{1}{2} \left(\rho_K^\xi + i \rho_K^\eta \right) U(\mathbf{x}_K) \right] \quad (10\text{-}18)$$

as suggested by Fermi's equation (10-17). Here $U(\mathbf{x}_K)$ is the quantized nuclear force field, and it is a complex wave which satisfies the Klein-Gordon equation just like Pauli and Weisskopf's $\Psi(\mathbf{x})$, discussed in lecture 6. If we use the two real fields U^ξ and U^η such that

$$U = \frac{1}{2}(U^\xi - iU^\eta), \qquad U^\dagger = \frac{1}{2}(U^\xi + iU^\eta), \quad (10\text{-}19)$$

then (10-18) can be rewritten as

$$g \sum_K \frac{1}{2} \left[\rho_K^\xi U^\xi(\mathbf{x}_K) + \rho_K^\eta U^\eta(\mathbf{x}_K) \right], \quad (10\text{-}18')$$

and as you see, in both (10-18) and (10-18') isospin plays a role. Furthermore, the exchange force also acts between two protons and between two neutrons, and therefore it was necessary to postulate the existence of a meson which is

---

4. This particle was actually a muon which has a larger penetrability. The pion was discovered in cosmic rays by Lattes, Muirhead, Occhialini, and Powell in 1947 (Lattes C M G, Muirhead H, Occhialini G P S, and Powell C F 1947 Processes Involving Charged Mesons *Nature* **159** 694–698).

also electrically neutral. In order for this exchange force to be identical to the exchange force between a neutron and a proton in the ϕ_s state, we must use instead of (10-18')

$$g \sum_K \frac{1}{2} \left[\rho_K^\xi U^\xi(\mathbf{x}_K) + \rho_K^\eta U^\eta(\mathbf{x}_K) + \rho_K^\zeta U^\zeta(\mathbf{x}_K) \right] =$$

$$g \sum_k \frac{1}{2} \left(\boldsymbol{\rho}_K \cdot \mathbf{U}(\mathbf{x}_K) \right), \tag{10-18''}$$

where $U^\zeta(\mathbf{x}_K)$ is the field for a neutral meson. (Here $U^\zeta(\mathbf{x}_K)$ is real in order for the particle to be neutral.) In this last formula $\mathbf{U}$ is considered to be a vector in $\xi\eta\zeta$ space, and this allows us to think that the nuclear-force boson also has the new fifth degree of freedom. If we adopt this interpretation, then instead of an isospin of $1/2$ for nucleons, the isospin of the nuclear-force boson is 1. And the ζ-component of this isospin can take the three values $+1, 0, -1$, where $+1$ indicates a positive meson, 0 a neutral meson, and -1 a negative meson. This corresponds to the case of nucleons for which the ζ-component of isospin $+1/2$ signifies the neutron state and $-1/2$ signifies the proton state. If we examine (10-18''), we see further that $\xi\eta\zeta$ space is completely isotropic in the meson theory also. In this sense, this space and isospin, which is defined in this space, have acquired an even deeper fundamental meaning.

Yukawa published his idea on the origin of the nuclear force in 1935; for a while it did not attract attention. For example, Oppenheimer did not agree with Yukawa's view, even after Anderson's discovery. Meanwhile, Yukawa's group clarified and extended his idea, and by considering the vector field rather than the original scalar field, they attempted to explain the term containing $(\sigma_K \cdot \sigma_L)$ in the nuclear force and to explain the origin of the anomalous magnetic moment (neither of which had been explained using a scalar field) of nuclei and showed that they can be explained qualitatively.

In the meantime in Europe, ideas like Yukawa's occurred to several people in order to relate the particle discovered by Anderson to the nuclear force, and those people, without knowing Yukawa's work, began to work out a similar theory. They were surprised to learn that Yukawa had already published his ideas and that Yukawa and company were trying to do the same thing as they were doing. For them, it seems the surprise of discovering Yukawa was greater than the surprise of discovering the meson itself.

I was in the University of Leipzig (where Heisenberg and Hund were) from 1937 to 1939, and there I heard that until that time the *Proceedings of the Physico-Mathematical Society of Japan*, which were donated by Japan, were not read by anybody and were put into the book stacks immediately after their arrival. However, ever since Yukawa's paper appeared, the journal was displayed in the physics department library, and the pages of the papers by Yukawa and company were stained by readers' hands.

The reason why Yukawa's idea of the heavy quantum did not attract attention was partly because the journal was obscure, but also partly because Yukawa's idea was too far ahead of its time, just like Pauli's concept of the neutrino. Indeed, Bohr came to Japan in 1937 and heard about Yukawa's work, but he apparently did not show much interest in the theory. Also, Heisenberg, who in his paper referred to the idea that behind the exchange force is a shuttling boson electron with spin zero, proceeded in the direction which was away from that very idea. It seems to me, after considering all of these, that the reason for this rejection is that although one wall after another was being removed and although there were many new discoveries, there remained the stubborn prejudice that there was a sanctuary inside nuclei, and the physics community was allergic to new particles. I might note that the workplaces of Heisenberg, Fermi, and Yukawa, who successively removed the walls of the sanctuary, get farther and farther from Copenhagen where Bohr resided. This might mean that Bohr's influence gets weaker as you move farther away.

Nevertheless, we cannot yet say that the nuclear sanctuary has been completely eliminated by the progress from Heisenberg to Fermi to Yukawa. There still is the remaining sanctuary, i.e., the difficulty of the divergence which does not only pertain to the inside of nuclei but also outside, clinging to all elementary particles.[5] This wall of the sanctuary is infinitely high and is still forbidding the entrance of people even today.

These were the developments behind Pauli's theory of spin and statistics.

I hope that I have delivered to you in this lecture the process by which attitudes that were up-to-date in 1930 started to change drastically around 1932, and towards 1940 the physicists' arena moved from atoms and molecules to nuclei and further to elementary particles, and the groundwork was laid to go very deeply into nature. Speaking of 1940, the Second World War erupted in Europe in 1939, the previous year, and Japan made war on the European powers in 1941. Therefore, from 1940, physicists in each country had to go through hardship for some time, and it was not until the end of the war that the theory of elementary particles budded, stretched its stem and leaves, and finally blossomed in full color.

---

5. Tomonaga is referring to the difficulty of infinite self-energies of particles, which he himself (and others) helped overcome, for all practical purposes, by the renormalization theory.

The Thomas Factor Revisited

Is Thomas' Theory Valid for Anomalous Magnetic Moments?

From the previous lecture you probably saw that the history of physics had its last climax before the war around 1940. I could end this long story at this climax, but I feel that it is necessary to present the homework which Jensen left for me in 1972, the year before last, when he came to Japan.

As I told you in lecture 9, the anomalous magnetic moment of the proton was discovered in 1933, and that of the neutron was determined in the following year. Today I would like to discuss what the Thomas factor would be for these anomalous magnetic moments. I sketched Thomas' idea in the second lecture. One day in the year before last, Jensen casually dropped by my house (casually describes his attire: wearing sandals and without a tie) and asked me whether he could start telling me a story from the history of physics. I said, "Go ahead!" and he said, "You must know the Thomas factor," and the story went on from there. According to Jensen, Thomas' theory made quite a splash in Europe, and there was a hot debate over its correctness. At any rate, from his theory, the factor of $1/2$ required by experiment is definitely explained. But, we must ascertain whether this factor of $1/2$ actually validates Thomas' theory or is simply a fortuitous agreement. If we compare the results from separately applying Thomas' idea and the new quantum mechanics to anomalous magnetic moments, then we shall know whether or not Thomas' idea is correct. This was Jensen's proposal when he came to my house.

By the way, the basis for the application of quantum mechanics to the anomalous magnetic moments had already been prepared by Pauli in 1933. Also in 1933 Stern first measured the anomalous magnetic moment of the proton, and in 1934 Rabi measured it more precisely. As I told you in lecture 9, when Pauli visited Stern's lab, he firmly believed that the magnetic moment of the proton was the normal value of $e\hbar/2m_p c$, and it is a little peculiar that Pauli had already published, as I shall tell you later, the tools necessary to attack the

Figure 11.1 Llewellyn H. Thomas (1903–1992).

problem of the anomalous value—which he did not believe—but in a way this is very much like Pauli.

It has already been more than half a year since I talked in lecture 2 about the Thomas factor so I repeat it here concisely in case you have forgotten. In the explanation of alkali doublet terms, instead of considering that an electron is revolving around the nucleus, we consider the coordinate system in which the electron is at rest and the nucleus is revolving around the electron. (I call this the electron rest system.) Then there will be a magnetic field at the electron's position caused by the nucleus revolving around the electron, and the magnetic moment of the electron surely interacts with the field, splitting the levels into doublets. This was the explanation of the alkali doublet terms. However, as I told you in lecture 2, if we do this, we get a value for the interval between doublet terms which is twice the experimental value.

Thomas solved this difficulty. According to him, the coordinate system in which the electron is at rest and the nucleus is revolving is not as simple as might naively be supposed. He noticed the necessity of carefully taking into account the special properties of relativity, especially those of the Lorentz transformation. If the electron is moving with a constant velocity, then the electron rest system is determined simply by a Lorentz transformation, but if the particle has acceleration, we must take into account a variety of things. This is the crux of Thomas' idea. I glossed over the detail of Thomas' calculation in lecture 2 because it was "too laborious," but we cannot answer Jensen's homework without it. Even though this may be a little bit complicated, please follow me. After all, it is not bad to review relativity.

Probably you already know the Lorentz transformation very well, but let me start from there.

Let us consider two arbitrary inertial frames I and I' and consider an orthogonal coordinate system S fixed on I and another orthogonal coordinate system S' fixed on I'; let the origin of S be O and its coordinate axes the X, Y, Z axes and the origin of S' be O' and its coordinate axes the X', Y', Z' axes. Let us examine some event from these two inertial frames, such that it is observed at time t at the point (x, y, z) in the I frame, and at time t' at coordinates (x', y', z') in the I' frame. Since in both the I frame and the I' frame the origin of coordinates and the zero point of time can be chosen arbitrarily, it is convenient to define them such that at the instant when $t = t' = 0$, O and O' are coincident. Then between the sets of variables $\{x, y, z, t\}$ and $\{x', y', z', t'\}$, there exists a homogeneous linear relation which satisfies

$$x^2 + y^2 + z^2 - c^2 t^2 = x'^2 + y'^2 + z'^2 - c^2 t'^2. \tag{11-1}$$

This equation (11-1) is a fundamental requirement of relativity, and when $\{x, y, z, t\}$ and $\{x', y', z', t'\}$ satisfy (11-1), then we say that these two sets of variables are related by a Lorentz transformation. Here (x, y, z) and (x', y', z') are "spatial" coordinates of the point where the event happened, and we correspondingly call $\{x, y, z, t\}$ and $\{x', y', z', t'\}$ the "space-time" coordinates at which the event happened in each inertial frame.

We can choose the coordinate systems S and S' on the inertial frames I and I' in a variety of ways, and accordingly we may have a variety of Lorentz transformations. It is therefore useful to separate them into nonrotational transformations and rotational ones.

Among the nonrotational Lorentz transformations, the well-known one with the simplest form is

$$x' = \frac{x - vt}{\sqrt{1 - v^2/c^2}}, \quad y' = y, \quad z' = z, \quad t' = \frac{t - \frac{v}{c^2}x}{\sqrt{1 - v^2/c^2}}. \tag{11-2}$$

If we solve these, we obtain the reverse transformation

$$x = \frac{x' + vt'}{\sqrt{1 - v^2/c^2}}, \quad y = y', \quad z = z', \quad t = \frac{t' + \frac{v}{c^2}x'}{\sqrt{1 - v^2/c^2}}. \tag{11-2'}$$

In order to consider the physical meaning of this transformation, we consider a fixed point P′ on the $I′$ frame and let its coordinates based on the $S′$ system be $(x'_{P'}, y'_{P'}, z'_{P'})$. We then see from (11-2) that its coordinates based on the S system viewed from the I frame are

$$x_{P'} - vt = \sqrt{1 - v^2/c^2}\, x'_{P'}, \quad y_{P'} = y'_{P'}, \quad z_{P'} = z'_{P'}.$$

These relations may be rewritten as

$$x_{P'} = \sqrt{1 - v^2/c^2}\, x'_{P'} + vt, \quad y_{P'} = y'_{P'}, \quad z_{P'} = z'_{P'},$$

and we see that the point P′ viewed from the I frame is at the position

$$(x_{P'}, y_{P'}, z_{P'}) = \left(\sqrt{1 - v^2/c^2}\, x'_{P'} + vt,\ y'_{P'},\ z'_{P'} \right) \tag{11-3}$$

at time t. Therefore, as seen from the I frame, point P′ is moving along the X-axis with a velocity $\mathbf{v}_{P'} = (v, 0, 0)$, and there is a Lorentz contraction with the factor $\sqrt{1 - v^2/c^2}$ along the X-axis. If as a special case we take P′ as the origin of the $S′$ system, then its coordinates seen from the I frame are

$$(x_{O'}, y_{O'}, z_{O'}) = (vt, 0, 0), \tag{11-4}$$

and it is also moving along the X-axis with a velocity

$$\mathbf{v}_{O'} = \mathbf{v}_{P'} = (v, 0, 0); \tag{11-5}$$

if we consider a vector O′P′, its components

$$(x_{P'} - x_{O'},\ y_{P'} - y_{O'},\ z_{P'} - z_{O'})$$

with respect to the S coordinates in the I frame are related to those of the $S′$ system by

$$\begin{aligned}
x_{P'} - x_{O'} &= \sqrt{1 - v^2/c^2}\,(x'_{P'} - x'_{O'}) \\
y_{P'} - y_{O'} &= y'_{P'} - y'_{O'} \\
z_{P'} - z_{O'} &= z'_{P'} - z'_{O'}.
\end{aligned} \tag{11-6}$$

Here also, the x-component is Lorentz-contracted.

Next let us take point P′ of (11-3) on the $X′$-axis. For this point, since $y'_{P'} = z'_{P'} = 0$, $y_{P'} = z_{P'} = 0$, and therefore if we look at the $X′$-axis in the I' frame from the I frame, then it is parallel to the X-axis. Similarly, if we look at the Y'-axis of the I' frame from the I frame, then it is parallel to the Y-axis; and likewise for the Z'-axis and the Z-axis.

So far we have examined how a point, a vector, and the coordinate axes fixed on the I' frame appear as seen from the I frame. Conversely, how do a point, a vector, and the coordinate axes fixed on the I frame appear from the I' frame? We can consider these similarly using (11-2'). The conclusion is that if we look at points fixed on the I frame from the I' frame, they all move along the X'-axis with velocity $-v$, and likewise vectors fixed on the I frame are moving parallel to the X'-axis with a velocity $-v$, and their x'-components are Lorentz-contracted by $\sqrt{1 - v^2/c^2}$. Furthermore, the coordinate axes X, Y, Z seen from the I' frame are parallel to the X', Y', Z' axes respectively. If we express these in formulas, we have

$$(x'_P, y'_P, z'_P) = \left(\sqrt{1 - v^2/c^2}\, x_P - vt', y_P, z_P \right) \qquad (11\text{-}3')$$

corresponding to (11-3),

$$(x'_O, t'_O, z'_O) = (-vt', 0, 0) \qquad (11\text{-}4')$$

corresponding to (11-4),

$$\mathbf{v}'_O = \mathbf{v}'_P = (-v, 0, 0) \qquad (11\text{-}5')$$

corresponding to (11-5), and

$$\begin{aligned} x'_P - x'_O &= \sqrt{1 - v^2/c^2}\, (x_P - x_O) \\ y'_P - y'_O &= y_P - y_O \\ z'_P - z'_O &= z_P - z_O \end{aligned} \qquad (11\text{-}6')$$

corresponding to (11-6).

Therefore, for the Lorentz transformation in the form of (11-2) or (11-2'), each of the three coordinate axes in the I frame is parallel to its counterpart in the I' frame. (This statement is a bit sloppy, and the real meaning of it is that, if you view other coordinate axes from one system, then they are parallel to the coordinate axes of its own, but it is usually convenient to be a bit sloppy.) In this sense, this transformation certainly does not include rotation of the coordinate axes and therefore is a nonrotational transformation. The transformation with coordinate axes parallel is also realized by the cyclic permutation of x, y, z in

(11-2). Since these transformations are special ones with the simplest form, they are often called *special Lorentz transformations*.

Since in a special Lorentz transformation the X-axis and X'-axis, the Y-axis and Y'-axis, and the Z-axis and Z'-axis are parallel to each other, one might think that if the components (a_x, a_y, a_z) in the S system of a vector $\mathbf{a}$ in the I frame and the components (a'_x, a'_y, a'_z) in the S' system of the vector $\mathbf{a}'$ in the I' frame have the relation among them $a_x : a_y : a_z = a'_x : a'_y : a'_z$, then $\mathbf{a}$ and $\mathbf{a}'$ are parallel. (This is also a sloppy way of saying it.) However, this is not the case. For as is clear from (11-6'), if we view $\mathbf{a}$ from the I' frame, its components are $(\sqrt{1 - v^2/c^2}\, a_x, a_y, a_z)$, and the vector (a'_x, a'_y, a'_z) in the I' frame which satisfies the relation $a'_x : a'_y : a'_z = a_x : a_y : a_z$ are never parallel unless either $a_x = a'_x = 0$ or $a_y = a'_y = a_z = a'_z = 0$. However when the coordinate axes of S and S' are parallel, it very often occurs that the vector $\mathbf{a}$ in the I frame and the vector $\mathbf{a}'$ in the I' frame satisfy $a_x : a_y : a_z = a'_x : a'_y : a'_z$, and they play a special role, so it is convenient to say they are "quasi-parallel" to each other. In general, a Lorentz transformation becomes a Galilean transformation in the limit that $\sqrt{1 - v^2/c^2}$ approaches 1, and in this limit, of course, quasi-parallel is the same as parallel.

I told you earlier that a special Lorentz transformation is nonrotational, but let us proceed to a more general nonrotational transformation. This transformation is realized when instead of the orthogonal coordinate systems S and S' taken on the I and I' frames (these are the coordinate systems for which the X and X' axes are parallel, the Y and Y' axes are parallel, and the Z and Z' axes are parallel), we take new orthogonal coordinate systems $\overline{S}$ and $\overline{S}'$ such that the $\overline{X}$ and $\overline{X}'$ axes are quasi-parallel, the $\overline{Y}$ and $\overline{Y}'$ axes are quasi-parallel, and the $\overline{Z}$ and $\overline{Z}'$ axes are quasi-parallel. According to the definition of quasi-parallelism which I mentioned earlier, if the $\overline{X}$ axis and the $\overline{X}'$ axis are quasi-parallel, that means that the direction cosines of the $\overline{X}$ axis with respect to the X, Y, Z axes and the direction cosines of the $\overline{X}'$ axis with respect to the X', Y', Z' axes are equal, and the same can be said for the $\overline{Y}$ axis and the $\overline{Y}'$ axis and for the $\overline{Z}$ axis and the $\overline{Z}'$ axis. Therefore, we can obtain the axes for the $\overline{S}$ and $\overline{S}'$ systems by making the same rotation of the axes of the S system and the S' system. If we take the space-time coordinates of some event in the $\overline{S}$ system as $\{\overline{x}, \overline{y}, \overline{z}, \overline{t}\}$ and in the $\overline{S}'$ system as $\{\overline{x}', \overline{y}', \overline{z}', \overline{t}'\}$, then we call the transformation between them $\{\overline{x}, \overline{y}, \overline{z}, \overline{t}\} \rightleftarrows \{\overline{x}', \overline{y}', \overline{z}', \overline{t}'\}$ a nonrotational Lorentz transformation. In this case the $\overline{X}, \overline{Y}, \overline{Z}$ axes of the $\overline{S}$ system and the $\overline{X}', \overline{Y}', \overline{Z}'$ axes of the $\overline{S}'$ system are quasi-parallel, and in this sense we can consider $\overline{S}$ and $\overline{S}'$ to be nonrotationally related. Needless to say, in the limit in which we can regard $\sqrt{1 - v^2/c^2}$ as 1, $\overline{S}$ and $\overline{S}'$ are literally parallel.

Furthermore, if the axes of the $\overline{S}$ system and the $\overline{S}'$ system are quasi-parallel and if the vectors $\overline{\mathbf{a}} = (\overline{a}_x, \overline{a}_y, \overline{a}_z)$ in the $\overline{S}$ system and $\overline{\mathbf{a}}' = (\overline{a}'_x, \overline{a}'_y, \overline{a}'_z)$ of the $\overline{S}'$ system satisfy the relation $\overline{a}_x : \overline{a}_y : \overline{a}_z = \overline{a}'_x : \overline{a}'_y : \overline{a}'_z$, then $\overline{\mathbf{a}}$ and $\overline{\mathbf{a}}'$ are quasi-parallel. We omit the proof because it is trivial. For these reasons, the quasi-parallelism of $\overline{\mathbf{a}}'$ and $\overline{\mathbf{a}}'$ can be defined without using coordinate systems which are parallel to each other; it is a concept which is independent of the choice of the $\overline{S}$ and the $\overline{S}'$ systems, and therefore it is invariant with respect to the transformation of coordinate systems.

What then is the general form of a nonrotational Lorentz transformation? If we do this for the general form, the story gets long and complicated with the risk of losing the essence of the problem. Therefore, we shall just generalize it to the degree necessary for later discussion. As I told you, the axes in the $\overline{S}$ system and those of the $\overline{S}'$ system are obtained by making the same rotation of the axes of S and S' used in (11-2) and (11-2'). Therefore, we consider the rotation of the X and X' axes and the Y and Y' axes around the Z and Z' axes by an angle α and make it the $\overline{S}$ system and the $\overline{S}'$ system. See figure 11.2. Then the point given by coordinates (x, y, z) in the S system is given by the spatial coordinates

$$\overline{x} = x \cos\alpha + y \sin\alpha$$
$$\overline{y} = -x \sin\alpha + y \cos\alpha \qquad (11\text{-}7)$$
$$\overline{z} = z$$

in the $\overline{S}$ system, and the point (x', y', z') in the S' system is given by

$$\overline{x}' = x' \cos\alpha + y' \sin\alpha$$
$$\overline{y}' = -x' \sin\alpha + y' \cos\alpha \qquad (11\text{-}7')$$
$$\overline{z}' = z'$$

in the $\overline{S}'$ system. (Needless to say, $\overline{t} = t$ and $\overline{t}' = t'$.) On the other hand, (x, y, z) and (x', y', z') are related by (11-2) and (11-2'), so after a little calculation we get

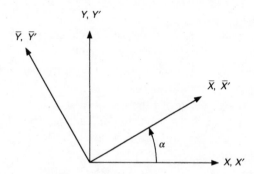

Figure 11.2

$$\bar{x}' = \left\{ \frac{1}{\sqrt{1 - v^2/c^2}} \cos^2 \alpha + \sin^2 \alpha \right\} \bar{x} -$$

$$\left\{ \frac{1}{\sqrt{1 - v^2/c^2}} - 1 \right\} \sin \alpha \cos \alpha \cdot \bar{y} - \frac{v \cos \alpha}{\sqrt{1 - v^2/c^2}} \cdot t$$

$$\bar{y}' = -\left\{ \frac{1}{\sqrt{1 - v^2/c^2}} - 1 \right\} \sin \alpha \cos \alpha \cdot \bar{x} + \qquad (11\text{-}8)$$

$$\left\{ \frac{1}{\sqrt{1 - v^2/c^2}} \sin^2 \alpha + \cos^2 \alpha \right\} \bar{y} + \frac{v \sin \alpha}{\sqrt{1 - v^2/c^2}} \cdot t$$

$$\bar{z}' = z' = \bar{z}$$

$$\bar{t}' = t' = \frac{t}{\sqrt{1 - v^2/c^2}} - \frac{\frac{v}{c^2}\{\cos \alpha \cdot \bar{x} - \sin \alpha \cdot \bar{y}\}}{\sqrt{1 - v^2/c^2}}$$

The reverse relations that give $\bar{x}, \bar{y}, \bar{z}, \bar{t}$ in terms of $\bar{x}', \bar{y}', \bar{z}', \bar{t}'$ can be obtained by exchanging $\bar{x}'$ and $\bar{x}$, $\bar{y}'$ and $\bar{y}$, $\bar{z}'$, and $\bar{z}$, and $\bar{t}'$ and $\bar{t}$, and replacing v with $-v$.

From the transformation (11-8) or from its inverse transformation, we immediately see that the coordinates of O' (O' is the origin of coordinates of the S' system, and it is at the same time the origin $\overline{O}'$ of the $\overline{S}'$ system) seen from the $\overline{S}$ system are

$$\bar{x}_{O'} = v \cos \alpha \cdot t, \quad \bar{y}_{O'} = -v \sin \alpha \cdot t, \quad \bar{z}_{O'} = 0, \qquad (11\text{-}9)$$

and the coordinates of O seen from the $\overline{S}'$ system are

$$\bar{x}'_O = -v \cos \alpha \cdot t', \quad \bar{y}'_O = v \sin \alpha \cdot t', \quad \bar{z}'_O = 0. \qquad (11\text{-}9')$$

Therefore the velocity of O' seen from the $\overline{S}$ system is

$$\mathbf{v}_{O'} = (v \cos \alpha, -v \sin \alpha, 0), \qquad (11\text{-}10)$$

and the velocity of O seen from the $\overline{S}'$ system is

$$\bar{\mathbf{v}}'_O = (-v \cos \alpha, v \sin \alpha, 0). \qquad (11\text{-}10')$$

Therefore, we finally have

$$\bar{\mathbf{v}}_{O'} = -\bar{\mathbf{v}}'_O \qquad (11\text{-}11)$$

and

$$|\bar{\mathbf{v}}_{O'}| = |\bar{\mathbf{v}}'_{O}| = |v|. \qquad (11\text{-}11')$$

It is a characteristic of the nonrotational Lorentz transformation that $\bar{\mathbf{v}}_{O'}$ and $\bar{\mathbf{v}}'_{O}$ are antiparallel. (In a special Lorentz transformation, in addition to this, $\bar{\mathbf{v}}_{O'}$ is along the direction of one of the X, Y, Z axes, and $\bar{\mathbf{v}}'_{O}$ is directly opposite to it.)

Finally let us consider the most general Lorentz transformation, namely the rotational Lorentz transformation. The rotational Lorentz transformation is a transformation in which the axes of the coordinate system of the I frame and the axes of the coordinate system on the I' frame are not quasi-parallel to each other. Let us take as their coordinate systems $\bar{\bar{S}}$ and $\bar{\bar{S}}'$, respectively. It is quite obvious that we can take $\bar{\bar{S}} = \bar{S}$ without loss of generality, and therefore we shall take it as such. We then see that the coordinate axes of the $\bar{S}$ system are $\bar{X}$, $\bar{Y}$, $\bar{Z}$, but the coordinate axes of $\bar{\bar{S}}'$ are given by some rotation of the coordinate axes $\bar{X}'$, $\bar{Y}'$, $\bar{Z}'$ of the $\bar{S}'$ system. Just as before, we shall limit ourselves to the rotation of the $\bar{S}'$ system around the $\bar{Z}'$ axis by an angle β. Then the coordinates of the point which were $(\bar{x}', \bar{y}', \bar{z}')$ in the $\bar{S}'$ system are given in the $\bar{\bar{S}}'$ system by $(\bar{\bar{x}}', \bar{\bar{y}}', \bar{\bar{z}}')$

$$\bar{\bar{x}}' = \cos\beta \cdot \bar{x}' + \sin\beta \cdot \bar{y}', \; \bar{\bar{y}}' = -\sin\beta \cdot \bar{x}' + \cos\beta \cdot \bar{y}', \; \bar{\bar{z}}' = \bar{z}'. \quad (11\text{-}12)$$

Obviously,

$$\bar{\bar{t}}' = \bar{t}'. \qquad (11\text{-}13)$$

Therefore, if we eliminate $\bar{x}'$, $\bar{y}'$, $\bar{z}'$, and $\bar{t}'$ from (11-12) and (11-8), then we obtain the formulas which express $\{\bar{\bar{x}}', \bar{\bar{y}}', \bar{\bar{z}}', \bar{\bar{t}}'\}$ in terms of $\{\bar{x}, \bar{y}, \bar{z}, \bar{t}\}$ and their inverses. Let us forget about writing these formulas because they will not be used explicitly. In fact, without using these formulas, we can determine the coordinates of the origin O of the $\bar{S}$ system as seen from the $\bar{\bar{S}}'$ system (which is the same as the origin of the $\bar{S}'$ system) and the origin O' of the $\bar{\bar{S}}'$ system seen from the $\bar{S}$ system. For example, the coordinates of O seen from the $\bar{S}'$ system are already given by (11-9'), and we immediately obtain

$$\bar{\bar{x}}'_{O} = -v\cos(\alpha + \beta) \cdot t', \quad \bar{\bar{y}}'_{O} = v\sin(\alpha + \beta) \cdot t', \quad \bar{\bar{z}}'_{O} = 0 \quad (11\text{-}14)$$

by substituting it into (11-12), and therefore seen from the $\bar{\bar{S}}'$ system, O is moving with the velocity

$$\bar{\bar{\mathbf{v}}}'_{O} = (-v\cos(\alpha + \beta), v\sin(\alpha + \beta), 0). \qquad (11\text{-}14')$$

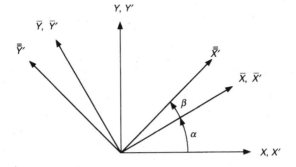

Figure 11.3

It is clear from figure 11.3 that the angle $\alpha + \beta$ appears in this formula. Furthermore for the origin O' of the $\bar{\bar{S}}'$ system seen from the S system, we can apply (11-9) and (11-10) directly, obtaining

$$\bar{x}_{O'} = v\cos\alpha \cdot t, \quad \bar{y}_{O'} = -v\sin\alpha \cdot t, \quad \bar{z}_{O'} = 0, \qquad (11\text{-}15)$$

and therefore it is moving with the velocity

$$\bar{\mathbf{v}}_{O'} = (v\cos\alpha, -v\sin\alpha, 0). \qquad (11\text{-}15')$$

Note in this case that

$$\bar{\mathbf{v}}_{O'} \neq -\bar{\bar{\mathbf{v}}}'_{O}. \qquad (11\text{-}16)$$

If we compare (11-14') and (11-15'), we find

$$-\bar{\bar{v}}'_{O_x} = \cos\beta \cdot \bar{v}_{O'_x} + \sin\beta \cdot \bar{v}_{O'_y} \qquad (11\text{-}17)$$
$$-\bar{\bar{v}}'_{O_y} = -\sin\beta \cdot \bar{v}_{O'_x} + \cos\beta \cdot \bar{v}_{O'_y}.$$

Since the coordinate axes of the $\bar{\bar{S}}'$ system are obtained by rotating those of the $\bar{S}$ system by an angle β, these results are well expected. Although $\bar{\mathbf{v}}_{O'} = -\bar{\bar{\mathbf{v}}}'_{O}$ is not followed in this case,

$$|\bar{\mathbf{v}}_{O'}| = |\bar{\bar{\mathbf{v}}}'_{O}| = |v| \qquad (11\text{-}17')$$

holds.

We now have three transformations: the special Lorentz transformation, the nonrotational Lorentz transformation, and the rotational Lorentz transformation. We have considered only rotations around the Z axis (and the Z' axis).

We can prove that all the Lorentz transformations are exhausted if we consider rotations around all axes. I told you that (11-16) is characteristic of rotational Lorentz transformations and (11-11) is characteristic of the nonrotational transformation. If we limit ourselves only to rotation about the Z axis (and the Z' axis), then we can say the converse, that is, that if (11-11) holds, then the transformation is nonrotational. However, for the case of the general rotation, even if (11-11) holds, still the transformation could be rotational. But let us not go into that here.

It took us a long time to review Lorentz transformations, but I thought that without this preparation, the crucial property of the Lorentz transformation which causes the Thomas factor could not be understood. Then what is it which causes the Thomas factor?

Let us consider three inertial frames I, I', I'' and consider the coordinate systems S on I, S' on I', and S'' on I''. Furthermore, let us assume that the axes of S and the axes of S' are quasi-parallel and that the axes of S' and the axes of S'' are also quasi-parallel. (I would like to remind you here that although we use a notation S, S', or S'' without bars for the coordinate system, this does not necessarily mean that S and S' or S' and S'' are related by a special Lorentz transformation.) In this case, are the coordinate axes of S and the coordinate axes of S'' mutually quasi-parallel? The answer to this question is, "Of course, yes," for the Galilean transformation. However, for a Lorentz transformation, the answer is, "It ain't necessarily so." In other words, even if the transformations $\{x, y, z, t\} \rightleftarrows \{x', y', z', t\}$ and $\{x', y', z', t'\} \rightleftarrows \{x'', y'', z'', t''\}$ are both nonrotational, in general the transformation $\{x, y, z, t\} \rightleftarrows \{x'', y'', z'', t''\}$ is rotational, except for the special case.

Since the general argument is too complicated, let us show this fact for the following simplified example. First, let the transformation $\{x, y, z, t\} \rightleftarrows \{x', y', z', t'\}$ be

$$x' = \frac{x - ut}{\sqrt{1 - u^2/c^2}}, \quad y' = y, \quad z' = z, \quad t' = \frac{t - \frac{u}{c^2}x}{\sqrt{1 - u^2/c^2}}. \quad (11\text{-}18)$$

In this case, the coordinate axes in the S system and in the S' system are not only quasi-parallel but also parallel. Next let us consider the transformation $\{x', y', z', t'\} \rightleftarrows \{x'', y'', z'', t''\}$.

$$x'' = x', \quad y'' = \frac{y' - vt'}{\sqrt{1 - v^2/c^2}}, \quad z'' = z', \quad t'' = \frac{t' - \frac{v}{c^2}y'}{\sqrt{1 - v^2/c^2}}. \quad (11\text{-}19)$$

In this case also, the axes in the S' system and those in the S'' system are not only quasi-parallel but also parallel. We can easily see by eliminating $\{x', y', z', t'\}$

from (11-18) and (11-19) that the transformation $\{x, y, z, t\} \rightleftarrows \{x'', y'', z'', t''\}$ is given by

$$x'' = \frac{x - ut}{\sqrt{1 - u^2/c^2}}, \quad y'' = \frac{\sqrt{1 - u^2/c^2} \cdot y + \frac{uv}{c^2}x - vt}{\sqrt{1 - u^2/c^2}\sqrt{1 - v^2/c^2}},$$

$$z'' = z, \quad t'' = \frac{t - \frac{u}{c^2}x - \frac{v}{c^2}\sqrt{1 - u^2/c^2} \cdot y}{\sqrt{1 - u^2/c^2}\sqrt{1 - v^2/c^2}}. \tag{11-20}$$

We immediately see from (11-20) that if we take $(x_{O''}, y_{O''}, z_{O''})$ as the coordinates of the origin O'' of the S'' system seen from the S system, then they are given by

$$x_{O''} = ut, \quad y_{O''} = \sqrt{1 - u^2/c^2}\, vt, \quad z_{O''} = 0, \tag{11-21}$$

and if we take (x_O'', y_O'', z_O'') as the coordinates of the origin O of the S system seen from the S'' system, then they are

$$x_O'' = -\sqrt{1 - v^2/c^2}\, ut'', \quad y_O'' = -vt'', \quad z_O'' = 0. \tag{11-21'}$$

Therefore, if we let the velocity of O'' seen from the S system be $\mathbf{w}_{O''}$ and the velocity of O seen from the S'' system be $\mathbf{w}_O''$, then they are

$$\mathbf{w}_{O''} = (u, \sqrt{1 - u^2/c^2}\, v, 0) \tag{11-22}$$

$$\mathbf{w}_O'' = (-\sqrt{1 - v^2/c^2}\, u, -v, 0). \tag{11-22'}$$

Obviously $\mathbf{w}_{O''} \neq -\mathbf{w}_O''$. However, their magnitudes are equal and are given by

$$|\mathbf{w}_{O''}| = |\mathbf{w}_O''| = \sqrt{u^2 - \frac{u^2 v^2}{c^2} + v^2}. \tag{11-23}$$

As I told you earlier, the fact that $\mathbf{w}_{O''} \neq -\mathbf{w}_O''$ clearly shows that the transformation $\{x, y, z, t\} \rightleftarrows \{x'', y'', z'', t''\}$ is rotational.

Then what is the rotation? In other words, by how much are the coordinate axes in the S'' system rotated from those of the S system. More rigorously, if we consider the $\overline{X}$-axis in the I frame, which is quasi-parallel to the X'' axis in the S'' system, and the $\overline{Y}$-axis in the I frame, which is quasi-parallel to Y'', then by how much is the coordinate system $\overline{S}$ composed of the $\overline{X}, \overline{Y}, \overline{Z}$ axes in the I

frame rotated from the original S system? In order to answer this question, we proceed as follows.

First of all, the axes in the $\overline{S}$ system considered in the I frame are quasi-parallel to those in the S'' system, and therefore the transformation $\{\overline{x}, \overline{y}, \overline{z}, \overline{t}\} \rightleftarrows \{x'', y'', z'', t''\}$ is nonrotational. [In this case according to (11-20), $z = z''$.) Therefore, between the velocity of O'' seen from $\overline{S}$ and the velocity of O seen from S'' is the relation (11-11). Thus if we take the velocity of O'' as viewed from $\overline{S}$ as $\overline{w}_{O''}$, then

$$\overline{w}_{O''} = -w''_O \qquad (11\text{-}24)$$

holds. On the other hand, as seen in figure 11.4, if the $\overline{S}$ system is rotated by an angle θ from the S system, then there are the relations between the components of $\overline{w}_{O''}$, i.e. $\overline{w}_{O''_x}$ and $\overline{w}_{O''_y}$, and the components of $w_{O''}$, i.e. $w_{O''_x}$ and $w_{O''_y}$,

$$\overline{w}_{O''_x} = w_{O''_x} \cos \theta + w_{O''_y} \sin \theta \qquad (11\text{-}25)$$
$$\overline{w}_{O''_y} = -w_{O''_x} \sin \theta + w_{O''_y} \cos \theta.$$

(In our case, it suffices to consider rotation only around the Z axis because $z = z''$.) Therefore, by expressing the left-hand side by the components of w''_O using (11-24), and using (11-22) and (11-22'), we obtain

$$\sqrt{1 - v^2/c^2}\, u = u \cos \theta + \sqrt{1 - u^2/c^2}\, v \sin \theta,$$
$$v = -u \sin \theta + \sqrt{1 - u^2/c^2}\, v \cos \theta \qquad (11\text{-}26)$$

From these equations we immediately have

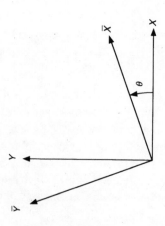

Figure 11.4

$$\cos\theta = \frac{\sqrt{1 - v^2/c^2} \cdot u^2 + \sqrt{1 - u^2/c^2} \cdot v^2}{u^2 - \frac{u^2 v^2}{c^2} + v^2}$$

$$\sin\theta = \frac{uv\left(\sqrt{1 - u^2/c^2}\,\sqrt{1 - v^2/c^2} - 1\right)}{u^2 - \frac{u^2 v^2}{c^2} + v^2}. \tag{11-27}$$

Finally we arrive at the conclusion that the coordinate axes of the S'' system are rotated by the angle θ given by (11-27) around the Z axis (= Z'' axis) of the S system. (More accurately, the coordinate axes of the $\overline{S}$ system, which are quasi-parallel to the axes of the S'' system are so rotated from the S system.) We have now seen that if we make a nonrotational transformation twice, the result becomes rotational. It was on this point that Thomas focused his attention.

Let us now go into Thomas' theory. Here Thomas introduced the concept of "proper coordinate axes of the particle." At the beginning of today's lecture, I said, "the coordinate system in which the electron is at rest but the nucleus is revolving"; I shall call this coordinate system in which the particle seems to be at rest at the origin the *rest system* of the particle. In general, if the particle is in uniform motion, then we can choose for its rest system the inertial system which is moving with the particle at uniform speed having coordinate axes such that the position of the particle is at their origin, and there is no problem. If, however, the velocity of the particle is not constant but changes with time, then we must take the inertial system which is moving with the particle's velocity at each instant. So far, so good, but how do we choose the coordinate axes in this inertial system? First of all, it is quite obvious to take the position of the particle as the origin of coordinates for each instant. But how do we choose the direction of the coordinate axes? Here a slightly complicated problem arises. If we take the direction of the coordinate axes for each instant arbitrarily, then there is no way to create a consistent theory. Therefore, Thomas proposed the following method.

First we consider a rest system at an arbitrary instant and another rest system at an infinitesimally later time. Thomas' idea is to take the axes of the latter such that they are quasi-parallel to the axes of the former. He then takes successive axes such that the axes of its rest system at one instant are quasi-parallel to the axes of the rest system at an infinitesimally earlier time. If we do it this way, once the motion of the particle is given and the initial axes are fixed, then the directions of the axes at any moment thereafter are successively, uniquely determined. When the axes of each instantaneous rest system are determined this way, they are called the *proper coordinate axes* of the particle.

Since it is very complicated to discuss the general situation, we shall consider a simple case where (11-18) and (11-19) and the various relations derived from them can be used. Let us consider the laboratory frame as the I frame and consider the coordinate system S fixed on it. Let us assume that the particle is instantaneously at the coordinate origin O at time $t = 0$. Furthermore, if the velocity of the particle at this time is **u**, then we take the X-axis along the direction of **u**. The Y and Z axes can be taken arbitrarily. We now consider the S' system, which is related to this laboratory system by Lorentz transformation (11-18), namely

$$x' = \frac{x - ut}{\sqrt{1 - u^2/c^2}}, \quad y' = y, \quad z' = z, \quad t' = \frac{t - \frac{u}{c^2}x}{\sqrt{1 - u^2/c^2}}. \quad (11\text{-}28)$$

In this system the origin O$'$ coincides with O at time $t = 0$, and the particle which is moving at this instant with velocity $\mathbf{u} = (u, 0, 0)$ in the S system is at rest instantaneously at O$'$ in the S' system. Therefore, it is obvious that the S' system is the rest system for the particle at time $t = 0$. It is a rest system in which the X', Y', and Z' axes are respectively parallel to the X, Y, and Z axes.

Next we consider the coordinate system S'' which is on the I'' inertial frame, whose meaning we consider later, and assume that the transformation between the S'' and S' systems is given by (11-19). We just use Δv instead of v for the convenience of later discussion. Therefore, the transformation is

$$x'' = x', \quad y'' = \frac{y' - \Delta v \cdot t'}{\sqrt{1 - (\Delta v)^2/c^2}},$$

$$z'' = z', \quad t'' = \frac{t' - \frac{\Delta v}{c^2}y'}{\sqrt{1 - (\Delta v)^2/c^2}}. \quad (11\text{-}29)$$

We now consider that, as seen from the laboratory system, the particle is at the origin O$''$ of the S'' system and is moving with the same speed at the instant $t = \Delta t$. This means that at the instant $t = \Delta t$, the S'' system is the rest system of the particle. We thus have given the meaning of S''. How then is the particle moving? In other words, as seen from the laboratory, where is the particle and with what speed is it moving at time $t = \Delta t$? Since we considered the particle to be moving with O$''$, it must have the position and velocity of O$''$ as seen from the laboratory system S.

Now recall (11-21). The coordinates $(x_{O''}, y_{O''}, z_{O''})$ of O$''$ seen from the S system are given by (11-21). We here need to use Δt instead of t, however. Then at this instant, the coordinates of the particle are given by

$$\mathbf{x} = (u\Delta t, \sqrt{1 - u^2/c^2}\,\Delta v \cdot \Delta t, 0). \quad (11\text{-}30)$$

As for the velocity of the particle, it is

$$\mathbf{w} = (u, \sqrt{1 - u^2/c^2}\ \Delta v, 0), \tag{11-31}$$

using (11-22). Also remembering that the velocity of the particle was $\mathbf{u} = (u, 0, 0)$ at $t = 0$, the particle has changed its velocity by $(0, \sqrt{1 - u^2/c^2} \cdot \Delta v, 0)$ perpendicularly to $\mathbf{u}$ during the time Δt. Therefore, if we make Δt infinitesimally small, the acceleration of the particle $\mathbf{a}$ is given by

$$\mathbf{a} = (0, a, 0), \quad a = \sqrt{1 - u^2/c^2}\ \frac{\Delta v}{\Delta t}, \quad \mathbf{a} \perp \mathbf{u}. \tag{11-32}$$

Because Δt is infinitesimally small, Δv is also infinitesimally small, and therefore we can ignore the v^2 term in (11-23) and obtain

$$|\mathbf{w}| = |\mathbf{u}| = |u|. \tag{11-33}$$

Therefore, the direction of the velocity of the particle changes between $t = 0$ and $t = \Delta t$, but its magnitude remains constant.

Now the transformation from the S' system to the S'' system is nonrotational. (In our case, the coordinate axes of the S' system and those of the S'' system were not only quasi-parallel but also parallel.) Therefore, the S'' system is not only the rest system for the particle at the instant $t = \Delta t$, but also its axes are the proper coordinate axes for the particle. Now by how much are the proper coordinate axes in the S'' system rotated with respect to the axes of the laboratory system S? The answer is given by (11-27). Here we write Δv for v and, correspondingly, $\Delta\theta$ for θ. Furthermore, since Δv is infinitesimally small, we can ignore the term with $(\Delta v)^2$. We then obtain

$$\Delta\theta = \frac{\Delta v}{u}\left(\sqrt{1 - u^2/c^2} - 1\right) \tag{11-34}$$

from (11-27), and using (11-32),

$$\Omega = \frac{\Delta\theta}{\Delta t} = -\left(\frac{1}{\sqrt{1 - u^2/c^2}} - 1\right)\frac{a}{u} \tag{11-35}$$

is derived. That is to say, the coordinate axes of the S'' system, namely the proper coordinate axes of the particle, are rotating with an angular velocity Ω around the Z-axis.

In deriving this conclusion we used for the rest system S' at the initial instant $t = 0$ coordinate axes that are parallel to the axes of the laboratory system S. [The transformation (11-28) was a special Lorentz transformation.] Instead of the coordinate axes of S', we could have started with those of $\overline{S}'$, which is rotated from S' by an angle α around the Z-axis, and the conclusion would have been the same. This is because in this case the proper coordinate axes of the $\overline{S}''$ system used at the instant $t = \Delta t$ are quasi-parallel to those of the $\overline{S}'$ system. Therefore, $\overline{S}''$ is obtained by rotating the axes of the S'' system by the angle α, and the relative relationship (e.g., the angle $\Delta\theta$ between them) between the axes of $\overline{S}'$ and $\overline{S}''$ as seen from the S system is identical to that between the S' and S'' axes. We can therefore regard the angular velocity of the proper coordinate axes to be independent of the proper coordinate axes which were assumed at the initial instant, and therefore the angular velocity of the proper coordinate axes is determined only by the velocity and acceleration of the particle, and it is an invariant quantity, independent of the choice of coordinate axes. In other words, it is an intrinsic quantity of the motion of the particle.

So far, we have examined two instants, $t = 0$ and $t = \Delta t$, but we can discuss the same thing for instants $t = \Delta t$ and $t = 2\Delta t$, $t = 2\Delta t$ and $t = 3\Delta t$, ... If in these cases we take the acceleration $\mathbf{a}$ to be perpendicular to the velocity $\mathbf{u}$ at every instant and the value of $|\mathbf{a}|$ to be constant, then the successive proper coordinate axes will continue to rotate according to (11-35). For this reason, the proper coordinate axes seen from the laboratory system will continue rotating with angular velocity (11-35) although from one instant to the next they are always taken to be quasi-parallel. Actually, the rotation given by (11-35) is not the rotation of the proper coordinate axes themselves seen from the laboratory system but the rotation of three vectors in the laboratory system assumed to be quasi-parallel to the proper coordinate axes. However, we can prove for the case $u^2/c^2 \ll 1$, such as for an electron in an atom, that we can regard (11-35) to be the rotation of the proper coordinate axes themselves. (I skip this proof.)

In this approximation, furthermore, we can use

$$\frac{1}{\sqrt{1 - u^2/c^2}} \approx 1 + \frac{u^2}{2c^2},$$

and obtain

$$\Omega = -\frac{ua}{2c^2}. \tag{11-35'}$$

Needless to say, the motion with $\mathbf{a} \perp \mathbf{u}$ and $a = $ constant at any instant is uniform circular motion, and its angular velocity is

$$\omega = \frac{a}{u}, \tag{11-36}$$

and its radius is

$$r = \frac{u^2}{a}. \tag{11-37}$$

Comparing with this ω, we get

$$\Omega = -\frac{1}{2}\frac{u^2}{c^2}\omega, \tag{11-35''}$$

and indeed the rotation of the proper coordinate axes is the outcome of relativity.

For simplicity we assumed above that the particle is executing uniform circular motion. We can show that in general, if the velocity and acceleration of the particle are given, then the angular velocity of the proper coordinate axes can be determined. According to Thomas' calculation, if the velocity is **u** and the acceleration is **a**, then the angular velocity is

$$\mathbf{\Omega} = -\frac{1}{2c^2}\mathbf{u} \times \mathbf{a}. \tag{11-38}$$

Our (11-35') is a special case of this, but for further discussion it is more convenient to use the general formula (11-38). [Formula (11-38) is nothing but (2-14) of lecture 2.]

Now we will finally go into the story of spin. First let us start from the classical motion of a top. As a concrete example, let us consider a ship cruising with a gyrocompass. In this case, if there is no torque operating on the top in the gyro, the rotational axis of the top is always pointing in one direction in space regardless of the acceleration with which the ship is moving. When I say a certain direction in space, I mean with respect to the coordinate axes fixed to the stars of the night sky. If there is a torque operating on the top, as you know, the top undergoes a neck-precession, the head of the top describing a circle.

Let us consider the spin of a particle from this analogy and do so relativistically. If we are to use relativity, we cannot assume the uniqueness of the coordinate axes fixed in the star system as in the classical theory. What shall we do? Thomas thought that he should use the proper coordinate axes which he introduced. Tentatively he considers that a particle with spin is rotating like a top. Now the motion of the particle may include acceleration in the laboratory system just like a gyrocompass on a ship. If there is no external torque on the rotational axis of the particle, then the spin of the particle (namely the rotational axis) always maintains a constant direction with respect to the proper coordinate axes. This was Thomas' assumption. This idea is a natural result of the analogy with the ordinary top, namely, just as in relativity the proper time intrinsic in the motion of the particle plays the role of absolute time in Newtonian mechanics, the proper coordinate axes intrinsic in the motion of the particle play the role of coordinate axes fixed in the star system.

What conclusion emerges from this idea? According to Thomas, if there is no torque on the particle and therefore if the spin maintains a constant direction with respect to the proper coordinate axes, then as seen from the laboratory system, it precesses with the angular velocity (11-38). This precession occurs without torque. This is the conclusion which Thomas drew, and therefore it is often called "Thomas precession."

Then what happens if there is a torque? For a classical top, we can actually think about such torque if we use a bar magnet with magnetic moment M (here we measure the magnetic moment in the ordinary units) for the axis of the top and spin it in a magnetic field $\mathbf{H}$. In this case, if the torque is not too large, the rotational axis of the spin starts to slowly precess around the magnetic field. Please look at figure 11.5. According to the classical mechanics of the top, the angular velocity of precession is given by

$$S\Omega_{\mathrm{H}} = -M\mathbf{H}. \qquad (11\text{-}39)$$

Here S is the angular momentum of the precession of the top (S is measured in the ordinary units).

Now in order to apply this classical conclusion to a particle with spin $1/2$, we should use for S and M

$$S = \frac{1}{2}\hbar \qquad M = \frac{-e}{2mc}gS. \qquad (11\text{-}40)$$

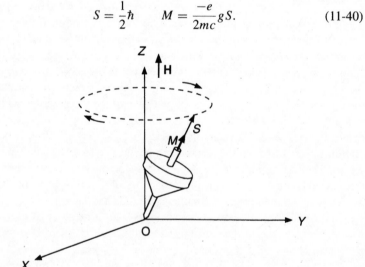

Figure 11.5

For an electron $g = g_0 = 2$, but for a proton $g \neq g_0$. We often call Dirac's theoretical value

$$M_D = \frac{-e}{2mc} g_0 S \qquad (11\text{-}40')$$

the normal spin magnetic moment, but for $g \neq g_0$ it is convenient to separate M into normal and abnormal parts like

$$M = M_D + M' \qquad (11\text{-}41)$$

(for the neutron $M_D = 0$). Needless to say, for the proton we must use the proton mass m_p for m on the right-hand side of the second formula of (11-40), and the sign of e must be changed. For today's lecture, however, we shall denote the mass as m without distinguishing the cases and always denote the charge as $-e$.

Thus even if we are using the classical model of a top for spin, we should keep in mind that (11-39) is valid with respect to the rest system of the particle and with respect to the proper coordinate axes. Since the proper coordinate axes themselves are rotating according to (11-38), the precession calculated from (11-39) is not quite the precession seen from the laboratory system. Namely, as a result of the rotation $\boldsymbol{\Omega}$ of the proper coordinate axes, the precession with angular velocity $\boldsymbol{\Omega}_H$ with respect to the proper coordinate axes will be precessing with the angular velocity

$$\boldsymbol{\Omega}_{lab} = \boldsymbol{\Omega}_H + \boldsymbol{\Omega} \qquad (11\text{-}42)$$

seen from the laboratory system. Furthermore, we must take into account that the magnetic field in the rest system is not only the external magnetic field $\mathbf{H}$ but also the added internal magnetic field $\mathring{\mathbf{H}}$ given by Biot-Savart's law. Therefore, (11-39) should take the form

$$S\boldsymbol{\Omega}_H = -M(\mathbf{H} + \mathring{\mathbf{H}}), \qquad (11\text{-}43)$$

and from (11-42)

$$S\boldsymbol{\Omega}_{lab} = -M(\mathbf{H} + \mathring{\mathbf{H}}) + S\boldsymbol{\Omega} \qquad (11\text{-}44)$$

is derived.

Now as for the acceleration of an electron in an atom, we have

$$\mathbf{a} = \frac{-e}{m} \mathbf{E}. \qquad (11\text{-}45)$$

(For the neutron, since $e = 0$,

$$\mathbf{a} = 0. \qquad (11\text{-}45)_{\text{neutron}}$$

We shall discuss this separately later.)

Substituting (11-45) into (11-38), we find

$$\boldsymbol{\Omega} = \frac{-e}{2mc^2}(\mathbf{E} \times \mathbf{u}). \qquad (11\text{-}46)$$

On the other hand, since as I told you in lecture 2, $\overset{\circ}{\mathbf{H}}$ is given by

$$\overset{\circ}{\mathbf{H}} = \frac{1}{c}(\mathbf{E} \times \mathbf{u}) \qquad (11\text{-}47)$$

[see (2-11) of lecture 2], we can write $\boldsymbol{\Omega}$ [referring to (11-40') with $g_0 = 2$] as

$$\boldsymbol{\Omega} = \frac{-e}{2mc}\overset{\circ}{\mathbf{H}} = \frac{1}{2S}M_{\text{D}}\overset{\circ}{\mathbf{H}}, \qquad (11\text{-}46')$$

and substituting this into (11-44), we obtain

$$S\boldsymbol{\Omega}_{\text{lab}} = -M\mathbf{H} - \left(M - \frac{1}{2}M_{\text{D}}\right)\overset{\circ}{\mathbf{H}}. \qquad (11\text{-}48)$$

From this final result we conclude the following. First for an electron, since $M = M_{\text{D}}$,

$$S\boldsymbol{\Omega}_{\text{lab}} = -M_{\text{D}}\mathbf{H} - \frac{1}{2}M_{\text{D}}\overset{\circ}{\mathbf{H}} \qquad (11\text{-}48)_{\text{electron}}$$

Lo and behold, there is this correct factor of $1/2$ in the second term on the right-hand side! This means that for the external field $\mathbf{H}$, the magnetic moment acts like M_{D}, but for the internal field $\overset{\circ}{\mathbf{H}}$, it acts apparently like only $(1/2)\,M_{\text{D}}$. This $1/2$ is *precisely* the Thomas factor. The reason for the appearance of this "apparent" $1/2$ is that, as shown by (11-46'), the kinematical quantity $\boldsymbol{\Omega}$ determined from the velocity and acceleration of a particle appears as if it came from the interaction of M_{D} and $\overset{\circ}{\mathbf{H}}$ in the case of a particle moving in the Coulomb field.

Next, for the proton, using (11-41), we obtain

$$S\boldsymbol{\Omega}_{\text{lab}} = -(M_{\text{D}} + M')\mathbf{H} - \left(\frac{1}{2}M_{\text{D}} + M'\right)\overset{\circ}{\mathbf{H}}. \qquad (11\text{-}48)_{\text{proton}}$$

Namely, the Thomas factor $1/2$ appears for the normal part M_{D} of M but does not appear for the abnormal part M'.

Finally for the neutron, since $\mathbf{a} = 0$, $\boldsymbol{\Omega} = 0$, and because $M_D = 0$, we obtain

$$S\boldsymbol{\Omega}_{\text{lab}} = -M'\mathbf{H} - M'\overset{\circ}{\mathbf{H}} \qquad (11\text{-}48)_{\text{neutron}}$$

from (11-48) and (11-41).[1] Therefore, in this case, the Thomas factor does not appear at all. By the way, the relations of (11-48) which are valid for arbitrary M are already in Thomas' paper. Therefore, he has already obtained the result for the anomalous magnetic moment also. Equation (11-44), which we discussed earlier, can be rewritten using (11-46), (11-40), and (11-47) as

$$\boldsymbol{\Omega}_{\text{lab}} = \frac{e}{2mc}g\mathbf{H} + \frac{e}{2mc}(g-1)\overset{\circ}{\mathbf{H}},$$

and if we put $g = g_0$, then $(11\text{-}48)_{\text{electron}}$ can be rewritten

$$\boldsymbol{\Omega}_{\text{lab}} = g_0\frac{e}{2mc}\mathbf{H} + (g_0 - 1)\frac{e}{2mc}\overset{\circ}{\mathbf{H}}.$$

This is (2-18') of lecture 2.

We have spent a lot of time on the classical theory of spin so let us now move on to the quantum theory.

At the beginning of this lecture, I told you that the quantum mechanical treatment of the anomalous magnetic moments had already been prepared by Pauli around 1933. This unfolded in the following way. As I told you in lecture 3, the question why the electron has spin angular momentum $\hbar/2$ and a magnetic moment $-e\hbar/2mc$ had been answered by the discovery of the Dirac equation; in the second edition of *Handbuch der Physik* published in 1933, Pauli discussed the Dirac equation in his article "The General Principles of Wave Mechanics." He pointed out that the Dirac equation is not the only first-order, linear equation which satisfies Lorentz invariance and other relativistic requirements. Namely, he noted that even if he adds a term, later called the "Pauli term", to the Dirac equation given in Lecture 3, the equation satisfies all the necessary requirements. However, if this term is added, the magnetic moment of a particle differs from $-e\hbar/2mc$. He terminated the discussion of this term with the following statement, which can be paraphrased: " . . . however, even if these terms are not added, the electron spin (or proton spin), as well as its magnetic moment $-e\hbar/2mc$, appears automatically. Hereafter, we shall proceed without this term." This *or proton spin* shows that Pauli prepared the manuscript for the *Handbuch* before Stern's experimental result was obtained, and what he wrote here is consistent with what he said to Stern.

---

1. Here we consider the neutron in atomic nuclei.

Figure 11.6 Otto Stern (1888–1969). [Courtesy of AIP Meggars Gallery of Nobel Laureates]

The Dirac equation in which the electric charge of a particle is $-e$ was already given in lecture 3, (3-20)

$$\left[\left(\frac{W}{c} + \frac{e}{c}A_0\right) - \sum_{r=1}^{3}\alpha_r\left(\frac{\hbar}{i}\frac{\partial}{\partial x_r} + \frac{e}{c}A_r\right) - \alpha_0 mc\right]\phi = 0. \quad (11\text{-}49)_D$$

Here we put $\psi = e^{-iWt/\hbar}\phi$ and study the eigenfunctions ϕ of the stationary state. Then as you know, W is the energy of this state. Here α_1, α_2, α_3, α_0 are the 4×4 matrices given by (3-23), and as I told you there, it is convenient to use the 2×2 Pauli matrices

$$\sigma_1 = \begin{pmatrix} 0 & 1 \\ 1 & 0 \end{pmatrix}, \quad \sigma_2 = \begin{pmatrix} 0 & -i \\ i & 0 \end{pmatrix}, \quad \sigma_3 = \begin{pmatrix} 1 & 0 \\ 0 & -1 \end{pmatrix} \quad (11\text{-}50)$$

and the 2×2 matrices

$$\mathbf{1} = \begin{pmatrix} 1 & 0 \\ 0 & 1 \end{pmatrix}, \quad \mathbf{0} = \begin{pmatrix} 0 & 0 \\ 0 & 0 \end{pmatrix} \quad (11\text{-}51)$$

to write the α_n as

$$\alpha_1 = \begin{pmatrix} 0 & \sigma_1 \\ \sigma_1 & 0 \end{pmatrix}, \quad \alpha_2 = \begin{pmatrix} 0 & \sigma_2 \\ \sigma_2 & 0 \end{pmatrix},$$

$$\alpha_3 = \begin{pmatrix} 0 & \sigma_3 \\ \sigma_3 & 0 \end{pmatrix}, \quad \alpha_0 = \begin{pmatrix} 1 & 0 \\ 0 & -1 \end{pmatrix}. \tag{11-52}$$

Also, it is convenient to write the four-component ϕ as composed of two-component quantities

$$\phi^+ = \begin{pmatrix} \phi_1 \\ \phi_2 \end{pmatrix}, \quad \phi^- = \begin{pmatrix} \phi_3 \\ \phi_4 \end{pmatrix}.$$

The Dirac equation $(11\text{-}49)_D$ is then rewritten in the form of the simultaneous equations

$$\{W + eA_0 - mc^2\}\phi^+ = c\sum_{r=1}^{3} \sigma_r \left(\frac{\hbar}{i} \frac{\partial}{\partial x_r} + \frac{e}{c} A_r \right) \phi^- \tag{$11\text{-}53)_{D_1}$}$$

$$\{W + eA_0 + mc^2\}\phi^- = c\sum_{r=1}^{3} \sigma_r \left(\frac{\hbar}{i} \frac{\partial}{\partial x_r} + \frac{e}{c} A_r \right) \phi^+. \tag{$11\text{-}53)_{D_2}$}$$

Pauli pointed out that even if a term of the form

$$-\frac{M'}{c}\left(\frac{1}{i}\sum_{\text{cyclic}} \alpha_0\alpha_2\alpha_3 H_1 - i\sum_r \alpha_0\alpha_r E_r \right) \tag{$11\text{-}49)_P$}$$

is added in the brackets of $(11\text{-}49)_D$, the equation satisfies the relativistic requirements. Here (H_1, H_2, H_3) are components of the magnetic field $\mathbf{H}$ acting on the particle, (E_1, E_2, E_3) are components of the electric field $\mathbf{E}$, and $\sum_{\text{cyclic}}$ means the addition of the terms over the cyclically permuted indices $(1, 2, 3)$, $(2, 3, 1)$, and $(3, 1, 2)$. The matrices in $(11\text{-}49)_P$ are the six-vectors discussed at the end of lecture 3, which are given explicitly as

$$\frac{1}{i}\alpha_0\alpha_2\alpha_3 = \begin{pmatrix} \sigma_1 & 0 \\ 0 & -\sigma_1 \end{pmatrix}, \quad \frac{1}{i}\alpha_0\alpha_3\alpha_1 = \begin{pmatrix} \sigma_2 & 0 \\ 0 & -\sigma_2 \end{pmatrix}, \quad \frac{1}{i}\alpha_0\alpha_1\alpha_2 = \begin{pmatrix} \sigma_3 & 0 \\ 0 & -\sigma_3 \end{pmatrix}$$

$$\tag{11-54}$$

$$i\alpha_0\alpha_1 = \begin{pmatrix} 0 & i\sigma_1 \\ -i\sigma_1 & 0 \end{pmatrix}, \quad i\alpha_0\alpha_2 = \begin{pmatrix} 0 & i\sigma_2 \\ -i\sigma_2 & 0 \end{pmatrix}, \quad i\alpha_0\alpha_3 = \begin{pmatrix} 0 & i\sigma_3 \\ -i\sigma_3 & 0 \end{pmatrix}.$$

The coefficient M' in (11-49)$_P$ is a constant which has the dimension of magnetic moment.

As I told you earlier, the electron spin $\hbar/2$, magnetic moment $-e\hbar/2mc$, and the Thomas factor $1/2$ all come out of the Dirac equation (11-53)$_D$, but Dirac's derivation is a little incomplete, so I would like to do it more rigorously. We adopt Pauli's method used in the *Handbuch*, mentioned earlier, and approximate (11-53)$_D$ semirelativistically. By "semirelativistic" I mean that since $W - mc^2$, eA_0, and eA_r are each smaller than mc^2, we can expand these quantities in powers of $1/c$, retain terms up to $1/c^2$, and neglect all the terms $1/c^3$, $1/c^4$, $\ldots$

In order to do this calculation, it is convenient to put

$$\left(\frac{\hbar}{i} \frac{\partial}{\partial x_r} + \frac{e}{c} A_r \right) \equiv \pi_r. \tag{11-55}$$

Then first from (11-53)$_{D_2}$ we obtain

$$\phi^- = \frac{c}{W + eA_0 + mc^2} \left(\sum_{r=1}^{3} \sigma_r \pi_r \right) \phi^+.$$

Rewriting

$$W + eA_0 + mc^2 = 2mc^2 + (W - mc^2) + eA_0,$$

considering that

$$W - mc^2 \ll mc^2, \qquad eA_0 \ll mc^2$$

and retaining up to the second-order terms of the approximation, we obtain

$$\phi^- = \left[\frac{1}{2mc} - \frac{W + eA_0 - mc^2}{(2mc)^2 c} \right] \left(\sum_{r=1}^{3} \sigma_r \pi_r \right) \phi^+.$$

Here the second term in brackets may appear negligible because it is a term with $1/c^3$, but we must retain this term for the reason you will see shortly. Now we substitute this expression for ϕ^- into the right-hand side of (11-53)$_{D_1}$. Then we have the equation satisfied by the two-component ϕ^+, and we shall see that this is the Pauli equation, which we discussed in lectures 3 and 7. Let us do this calculation.

First we shall perform the aforementioned substitution. Then we obtain

$$\text{RHS of (11-53)}_{D_1} = \left[\frac{1}{2m} \left(1 - \frac{W + eA_0 - mc^2}{2mc^2} \right) \sum_{r=1}^{3} \sigma_r \pi_r \sum_{r=1}^{3} \sigma_r \pi_r \right.$$

$$\left. - \frac{e}{(2mc)^2} \frac{\hbar}{i} \sum_{r=1}^{3} \sigma_r \frac{\partial A_0}{\partial x_r} \sum_{r=1}^{3} \sigma_r \pi_r \right] \phi^+ .$$

Indeed, the term containing $1/c^3$ has become a term with $1/c^2$ because of the factor c on the right-hand side of (11-53) $_{D_1}$. Thus we should not have neglected this.

Next we use the formula

$$\sum_{r=1}^{3} \sigma_r F_r \sum_{r=1}^{3} \sigma_r G_r = \sum_{r=1}^{3} F_r G_r + i \sum_{\text{cyclic}} \sigma_1 (F_2 G_3 - F_3 G_2), \quad (11\text{-}56)$$

which can be derived from the relations

$$\sigma_r^2 = 1, \quad \sigma_1 \sigma_2 = i \sigma_3, \quad \sigma_3 \sigma_1 = i \sigma_2, \quad \sigma_2 \sigma_3 = i \sigma_1$$

(here F_r and G_r could be q-numbers). We then have

$$\text{RHS of (11-53)}_{D_1} = \left[\frac{1}{2m} \left(1 - \frac{W + eA_0 - mc^2}{2mc^2} \right) \sum_{r=1}^{3} \pi_r \pi_r \right.$$

$$+ \frac{1}{2m} \left(1 - \frac{W + eA_0 - mc^2}{2mc^2} \right) i \sum_{\text{cyclic}} \sigma_1 (\pi_2 \pi_3 - \pi_3 \pi_2)$$

$$\left. - \frac{e}{(2mc)^2} \hbar \sum_{\text{cyclic}} \sigma_1 \left(\frac{\partial A_0}{\partial x_2} \pi_3 - \frac{\partial A_0}{\partial x_3} \pi_2 \right) - \frac{e}{(2mc)^2} \frac{\hbar}{i} \sum_{r=1}^{3} \frac{\partial A_0}{\partial x_r} \pi_r \right] \phi^+ .$$

We can easily see

$$\pi_2 \pi_3 - \pi_3 \pi_2 = \frac{e\hbar}{ic} \left(\frac{\partial A_3}{\partial x_2} - \frac{\partial A_2}{\partial x_3} \right) = \frac{e\hbar}{ic} H_1$$

(and the other relations in which 1, 2, 3 are cyclically permuted) and also note that

$$- \frac{\partial A_0}{\partial x_r} = E_r, \qquad r = 1, 2, 3.$$

Since in the classical theory for a particle with velocity $\mathbf{u} = (u_1, u_2, u_3)$ the momentum is

$$\pi_r \equiv \frac{mu_r}{\sqrt{1 - u^2/c^2}}, \qquad r = 1, 2, 3, \qquad (11\text{-}57)$$

we introduce the q-numbers u_r with the same relations, and neglecting the terms with $1/c^3$, we have

$$\text{RHS of (11-53)}_{D_1} = \left[\frac{m}{2} \frac{u^2}{1 - u^2/c^2} - \frac{W + eA_0 - mc^2}{4c^2} u^2 + \frac{e\hbar}{2mc} \sum_{r=1}^{3} \sigma_r H_r \right.$$

$$\left. + \frac{1}{2} \frac{e\hbar}{2mc} \sum_{\text{cyclic}} \sigma_1 \left(E_2 \frac{u_3}{c} - E_3 \frac{u_2}{c} \right) + \frac{i}{2} \frac{e\hbar}{2mc} \sum_{r=1}^{3} E_r \frac{u_r}{c} \right] \phi^+.$$

If we use this for (11-53)$_{D_1}$, the equation to be satisfied by ϕ^+ is determined to be

$$\left\{ \frac{m}{2} \left(1 + \frac{u^2}{c^2} \right) u^2 - M_D(\boldsymbol{\sigma} \cdot \mathbf{H}) - \frac{1}{2} M_D \left[\boldsymbol{\sigma} \cdot \left(\mathbf{E} \times \frac{\mathbf{u}}{c} \right) \right] - \frac{i}{2} M_D \left(\mathbf{E} \cdot \frac{\mathbf{u}}{c} \right) \right.$$

$$\left. - \left(1 + \frac{1}{4} \frac{u^2}{c^2} \right) (W + eA_0 - mc^2) \right\} \phi^+ = 0. \qquad (11\text{-}58)$$

Here $M_D = -e\hbar/2mc$. If we introduce the somewhat inconsistent approximation of equating to 1 the factor $(1 + u^2/c^2)$ of the first term in the braces and the factor $(1 + u^2/4c^2)$ of the last term in the braces (this corresponds to neglecting the relativistic change in the mass), we find that in addition to the first term in the braces $mu^2/2$ corresponding to the kinetic energy of the particle, and to $-eA_0$, the last term of $-(W + eA_0 - mc^2)$, corresponding to the potential energy of the particle due to the electric field, and the rest energy mc^2, additional terms appear—the second term, the third term, and the fourth term—that are added to the energy. The second term

$$- M_D(\boldsymbol{\sigma} \cdot \mathbf{H}) \qquad (11\text{-}59)_1$$

may be interpreted to be the interaction energy between the magnetic moment of the electron and the external magnetic field. Recalling (11-47), the third term is

$$- \frac{1}{2} M_D \left[\boldsymbol{\sigma} \cdot \left(\mathbf{E} \times \frac{\mathbf{u}}{c} \right) \right] = -\frac{1}{2} M_D(\boldsymbol{\sigma} \cdot \overset{\circ}{\mathbf{H}}), \qquad (11\text{-}59)_2$$

and this is the interaction energy between the magnetic moment of the electron and the internal magnetic field. Guess what? You see that the Thomas factor $1/2$ appears here just as in the classical Thomas theory of precession if you compare this with the right-hand side of $(11\text{-}48)_{\text{electron}}$. On the contrary, the interaction with an external magnetic field is given by $(11\text{-}59)_1$, and just as in $(11\text{-}48)_{\text{electron}}$, the factor $1/2$ does not appear there. The fourth term of $(11\text{-}58)$ does not exist in the classical theory, but this term does not contain the spin variable and therefore is not related to the Thomas factor. Therefore, we shall not discuss it further.

Here I would like to add some remarks on the first term $(m/2)\mathbf{u}^2$. From $(11\text{-}57)$ we have approximately

$$\frac{1}{2}m\mathbf{u}^2 = \frac{1}{2m}\mathbf{p}^2 + \frac{e}{2mc}(\mathbf{A}\cdot\mathbf{p} + \mathbf{p}\cdot\mathbf{A}),$$

but since for a homogeneous external field $\mathbf{A} = \mathbf{H}\times\mathbf{r}/2$, we obtain

$$\frac{1}{2}m\mathbf{u}^2 = \frac{\mathbf{p}^2}{2m} + \frac{e\hbar}{2mc}\mathbf{H}\cdot\boldsymbol{\ell}.$$

If we use this formula together with $(11\text{-}59)_1$ and $(11\text{-}59)_2$, the Pauli equation $(7\text{-}28)$ can be derived from $(11\text{-}58)$ without adding the Pauli term $(11\text{-}49)_P$.

What would happen if we add Pauli's additional term $(11\text{-}49)_P$? If we add it and use $(11\text{-}54)$, the equations are

$$\left(W + eA_0 - mc^2 + M'\sum_{r=1}^{3}\sigma_r H_r\right)\phi^+ =$$
$$c\sum_{r=1}^{3}\left(\sigma_r \pi_r + \frac{iM'}{c}\sigma_r E_r\right)\phi^-$$

$$(11\text{-}53)_{\text{D+P}}$$

$$\left(W + eA_0 + mc^2 - M'\sum_{r=1}^{3}\sigma_r H_r\right)\phi^- =$$
$$c\sum_{r=1}^{3}\left(\sigma_r \pi_r + \frac{iM'}{c}\sigma_r E_r\right)\phi^+.$$

Here M' is on the order of $1/c$. Therefore, we can neglect the term $M'\sum\sigma_r H_r$ on the left-hand side of the second equation. In this approximation, therefore, we can use

$$\phi^- = \left[\frac{1}{2mc} - \frac{W + eA_0 - mc^2}{(2mc)^2 c} \right] \sum_{r=1}^{3} \left(\sigma_r \pi_r - \frac{iM'}{c} \sigma_r E_r \right) \phi^+.$$

If we use this on the right-hand side of the first formula and calculate again using (11-56), we obtain many terms, but after neglecting terms of order $1/c^3, 1/c^4, \ldots$, only the terms

$$-\frac{M'}{2mc} \sum_{\text{cyclic}} \sigma_1 (E_2 \pi_3 - \pi_2 E_3 + \pi_3 E_2 - E_3 \pi_2) +$$

$$\frac{iM'}{2mc} \sum_{\text{cyclic}} (E_r \pi_r - \pi_r E_r) \qquad (11\text{-}60)$$

remain in addition to $(11\text{-}59)_1$ and $(11\text{-}59)_2$ for the case of $(11\text{-}53)_D$. If we use

$$F \pi_r - \pi_r F = -\frac{\hbar}{i} \frac{\partial F}{\partial x_r},$$

then

$$\text{first term of } (11\text{-}60) \ = -\frac{M'}{mc} \sum_{\text{cyclic}} \sigma_1 (E_2 \pi_3 - E_3 \pi_2) -$$

$$\frac{M' \hbar}{2imc} \sum_{\text{cyclic}} \sigma_1 \left(\frac{\partial E_2}{\partial x_3} - \frac{\partial E_3}{\partial x_2} \right)$$

and for the second term we obtain

$$\text{second term of } (11\text{-}60) \ = -\frac{M' \hbar}{2mc} \nabla \cdot \mathbf{E}.$$

However, this second term does not contain spin variables and therefore is not related to the Thomas factor, and we shall not discuss this term further. On the other hand, for the first term, remembering that $\mathbf{E}$ and $\mathbf{H}$ independent of time satisfy $\nabla \times \mathbf{E} = 0$, the additional term containing spin resulting from M' is only

$$-\frac{M'}{mc} \sum_{\text{cyclic}} \sigma_1 (E_2 \pi_3 - E_3 \pi_2).$$

Using $\pi_r = mu_r$ and noticing (11-47), we find

Figure 11.7 J. Hans D. Jensen,
1972. (1907–1973)

$$- M' \left[\boldsymbol{\sigma} \cdot \left(\mathbf{E} \times \frac{\mathbf{u}}{c} \right) \right] = - M'(\boldsymbol{\sigma} \cdot \mathring{\mathbf{H}}). \qquad (11\text{-}61)$$

Furthermore, on the left-hand side of $(11\text{-}53)_{D+P}$, there is an additional term

$$M'(\boldsymbol{\sigma} \cdot \mathbf{H}) \qquad\qquad (11\text{-}62)$$

which did not exist before, and so all together, the additional terms related to spin in the equation to be satisfied by ϕ^+ are, combining $(11\text{-}59)_1$, $(11\text{-}59)_2$, $(11\text{-}61)$, and $(11\text{-}62)$,

$$- (M_{\mathrm{D}} + M')(\boldsymbol{\sigma} \cdot \mathbf{H}) - \left(\frac{1}{2} M_{\mathrm{D}} + M' \right)(\boldsymbol{\sigma} \cdot \mathring{\mathbf{H}}). \qquad (11\text{-}63)_{\text{proton}}$$

If we compare this with the result $(11\text{-}48)_{\text{proton}}$ obtained classically using

Thomas' idea, we find they are exactly analogous. For the neutron, $e = 0$ and therefore $M_D = 0$, so the additional quantum mechanical term is $(11\text{-}63)_{\text{proton}}$ with $M_D = 0$. This is

$$- M'(\sigma \cdot \mathbf{H}) - M'(\sigma \cdot \mathring{\mathbf{H}}), \qquad (11\text{-}63)_{\text{neutron}}$$

and this is again analogous to $(11\text{-}48)_{\text{neutron}}$ obtained from Thomas' theory. Furthermore, the result obtained by Thomas' theory and that obtained by quantum theory not only look the same but can be shown to be completely identical if we translate it into the former by the correspondence principle. Therefore, we obtain an affirmative answer to the problem posed by Jensen. Namely, the answer obtained by Thomas' method and by quantum mechanics agree completely even for the anomalous magnetic moment.

My impression when I saw Jensen was that he seemed to be anticipating that the Thomas theory would not give the right answer for the anomalous magnetic moment. Now it is impossible to confirm this. I was thinking of writing to him, but lamentably he became ill and passed away in the spring of 1973. However, because of him I had a chance to look at Thomas' work again from the beginning. I told you earlier that Pauli withdrew his idea that spin is classically indescribable after learning of Thomas' work; Thomas' work indeed has sufficient content to reform Pauli.

My story has evolved and again come back to Thomas' time. Having completed this full circle, I would like to stop my talk. After today's long talk, which might have been a little bit dry, I prepare only one free-flowing talk. See you later, alligator.

The Last Lecture

Addenda and Recollections

I promised to give you *seidan* today. The meaning of *seidan*, as you probably know, originates from the seven wise men in the bamboo grove in the Weih period of China. These seven men rebelled against the stiff Confucianism, forsook the world, shut themselves up in the bamboo grove, played musical instruments and drank sake, forgot themselves amid the beauty of mountains and water, wrote poems, and discussed the philosophy of Lao-tse and Chuang-tze—this is called *seidan*. I would not dare imitate these seven wise men today, but what I meant is that for this last lecture we leave behind the rigorous discourse, and as professors often do in their last lecture, complement the material with anecdotes and recollections which come to mind. It is my wish that through such unassuming recollections I can sketch somewhat the status of physics in Japan from 1925 to 1940, which has been the stage for my lectures.

Actually, when I finished the first lecture of this series, Mr. Ishikawa of *Shizen*[1] brought me a book entitled *Theoretical Physics in the Twentieth Century*. This book was originally planned to celebrate Pauli's sixtieth birthday, but it was fated to be a memorial tribute; Mr. Ishikawa brought it to me because there are many stories about spin in it. I was afraid that if I had read it, then I might find that somebody else had already written about what I wanted to say. Therefore, for some time at the beginning of these lectures, I purposely did not read this book. After browsing through it much later, I was relieved to find that the talks which I gave based on my memories were not so wrong.

One story I have told is about Kronig. It is the episode in which Kronig thought of the self-rotating electron, talked to Pauli about it, and was categorically

---

1. A Japanese monthly journal published by Chuokoron-sha, Inc., now discontinued. *Shizen* means "nature."

rejected by Pauli. I think I heard this story when I came to RIKEN[2] from Professor Yoshikatsu Sugiura, who came back from Europe shortly before Professor Yoshio Nishina. Kronig himself describes this episode in the memorial book for Pauli, which I just mentioned. Therefore, in lecture 2 I have borrowed from his article.

Next is a story of Uhlenbeck and Goudsmit. About one year after Kronig had decided not to publish his idea because of the objections of Pauli and the Copenhagen group, Goudsmit and Uhlenbeck arrived at exactly the same idea of a self-rotating electron, and they managed to submit it to a journal. They also received various criticisms, and consequently they wanted to withdraw their paper. I told you in lecture 2 that by then the paper was already in the hands of the publisher. I heard this episode from R. E. Peierls, but recently I received material describing this story in detail from Professor Isamu Nitta. In *Delta*, a Dutch English-language journal, is a recollection by Goudsmit.

According to this recollection, Goudsmit was studying under Ehrenfest. Probably Ehrenfest sensed that Goudsmit was a better experimentalist than theorist, so he recommended Goudsmit to see Paschen at Bonn. Also in Paschen's lab was Back, and there they discovered the Paschen-Back effect, as well as experimentally confirming the hydrogen fine structure calculated by Sommerfeld. Goudsmit learned of a variety of experimental results, and while talking with these experimentalists, he was led to the idea of explaining the alkali doublet term by the coupling of the vectors l and s based on the idea of the fourth quantum number which Pauli had introduced in relation to his exclusion principle (of course he did not know that Kronig had the same idea). In addition, one of the very important facts he learned in Paschen's lab was the fine structure of H. These spectral lines which should be forbidden according to Sommerfeld's theory were definitely observed in Paschen's experiment. Goudsmit realized that if he used this new interpretation (this was also considered by Kronig) to consider the fine structure of H as similar to the alkali doublet term, then these lines are not forbidden but are naturally expected.

Goudsmit, however, was not a very capable theoretician, as he himself admits, and ideas such as the fourth degree of freedom and the self-rotating electron did not occur to him. However, he is very proud that he could explain that these forbidden lines are actually not forbidden.

Therefore, Goudsmit summarized these ideas in a paper and sent it to Copenhagen for the opinions of Kronig and Kramers. Goudsmit notes in his recollection that although a long letter came back from Kronig in which much was written on a variety of subjects, not a word was mentioned on Goudsmit's idea, which suggested that Kronig had absolutely no interest in it. Thus Kronig completely ignored Goudsmit's idea and danced around in the letter to different

---

2. *Rikagaku Kenkyujo* Institute of Physical and Chemical Research, *the* government-supported research institution in Japan, where scientists conducted research with utmost freedom.

subjects, and it seems that he was inflicting on Goudsmit the rejection he had incurred from Pauli. It seems to be a universal truth that a bride who is bullied by her mother-in-law will herself become a bad mother-in-law.

Uhlenbeck seems to have joined the battle around this time. He was Dutch and happened to be studying in Italy. There he was educated in classical physics but did not learn at all about the new quantum physics or spectroscopy. He returned to the Netherlands and joined Ehrenfest's group, whereupon Ehrenfest arranged for him to work with Goudsmit. According to Goudsmit, he was called to see Ehrenfest one day and was told, "Please work with Uhlenbeck for a while; then he will learn a lot from you about the new atomic structure and spectra." Goudsmit says that although Ehrenfest said it with a straight face, he must have been thinking, "Then you may learn from Uhlenbeck what the *real* physics is." Anyhow, Ehrenfest seems to have been an excellent educator who recognized the talent in his disciples, complemented their shortcomings, and guided them with very kind humanity so that each developed his originality. (Ehrenfest was modest, always saying that he was not a scholar but only a bush-league teacher. It seems to me that he did not say this out of modesty, but he genuinely thought that way, and I think it is not unrelated to the love he showed toward immature students. When Lorentz retired, Ehrenfest succeeded him, but only after he had turned down the offer for the position many times, saying that he was not worthy. He eventually accepted it after being persuaded by Lorentz, but throughout his life he had never forsaken the idea that he was not really qualified, and he ended his life by the tragic act of suicide.)

Pauli called the fourth quantum number "classically indescribable two-valuedness" and rejected all mechanical imagery, but Uhlenbeck thought this was the manifestation of the fourth degree of freedom of the electron and, more concretely, a self-rotation of the electron. Goudsmit could not quite understand the idea that the electron self-rotates but somehow liked it, and the two wrote it up in a joint paper and handed it to Ehrenfest. In the meantime, Uhlenbeck, who was versed in classical theory, thought that they should seek Lorentz's opinion, and therefore he saw Lorentz and presented their idea of the self-rotating electron. Lorentz said, "This idea is very difficult. Because if we adopt this idea, the magnetic self-energy becomes very large, and the mass of the electron is larger than that of the proton." Upon hearing this, Uhlenbeck got cold feet and went to Ehrenfest and said, "Please do not submit that paper. It probably is wrong." Ehrenfest said, "It is too late. That has been sent already." Goudsmit, on the other hand, was not considering it seriously anyhow, so he never dreamed that the idea of a self-rotating electron was wrong. Goudsmit remembers the words of Ehrenfest when they handed him the paper. "This is a good idea. Your idea may be wrong, but since both of you are so young without any reputation, you would not lose anything by making a stupid mistake." Thus the paper of the two lads went out into the world and was published.

This paper was submitted by Ehrenfest to the editor of *Naturwissenschaften* and was published, and a while after this paper was published, Uhlenbeck and Goudsmit sent another paper to *Nature*, which was harshly criticized by Kronig, as I told you in lecture 2. In the aforementioned *Theoretical Physics in the Twentieth Century*, Kronig had written an article which sounds a little bit like an excuse. He says since he was very coolly received when he voiced his idea in Copenhagen, yet within one year after that Bohr and all his minions there had made an about-face, all Kronig could do was simply call attention to the fact that there are certain points which fail even if the electron is considered to be self-rotating.

There were many other responses to this paper. For example, Heisenberg immediately wrote a letter to Goudsmit. He starts with "I have read your 'courageous' paper," and after writing a formula he asks the question, "How did you get rid of the factor of 2?" Upon reading this letter, Goudsmit had absolutely no idea either why their paper was so "courageous" or what was meant by the factor of 2. On the other hand, Uhlenbeck seems to have realized the difficulty of the factor of 2, in addition to the difficulty pointed out by Lorentz, and they clearly recognized these points in their paper in *Nature*. Uhlenbeck went so far as to try to withdraw their paper; he must have known very well that it was indeed a very "courageous" paper.

While this was going on, Thomas' paper was published, and for the time being, the difficulty of the factor of 2 was resolved. It was lucky for Goudsmit and Uhlenbeck that Ehrenfest, unlike Landé, did not advise them to consult Pauli. I shall quote from a photocopy of a letter from Thomas to Goudsmit which I found among the materials received from Professor Nitta.

> I think you and Uhlenbeck have been very lucky to get your spinning electron published and talked about before Pauli heard of it . . . more than a year ago, Kronig believed in the spinning electron and worked out something; the first person he showed it to was Pauli—Pauli ridiculed the whole thing so much that the first person became also the last and no one else heard anything of it.

Until Thomas' paper was published, Pauli was dogmatically opposed to the idea of the self-rotating electron. While Heisenberg was expressing his objection in a half-cynical, soft tone, such as " 'courageous' paper" or "How did you get rid of the factor of 2?" Pauli was relentlessly barking against Bohr's endorsement of Uhlenbeck and Goudsmit's paper. Namely, Bohr composed a note praising the paper by those two in *Nature*, but Pauli deplored this, saying that such an act by a person none other than Bohr simply introduced into atomic physics a new heretical doctrine. However, as I told you earlier, when Thomas' paper in which the discrepancy of the factor of 2 between experiment and theory was eliminated was published and the correct spacing of alkali doublets was clearly derived by the classical Thomas theory, Pauli could no longer maintain the objection "classically indescribable." Goudsmit says that soon after Thomas' paper was out, he received a postcard from Pauli saying, "I now believe in the idea of the self-rotating electron."

As you have seen here, the history of the discovery of spin has followed an interesting path showing a display of human relations among a variety of strong characters. In *Theoretical Physics in the Twentieth Century*, van der Waerden also writes in detail about Pauli's dealings with Kronig, Uhlenbeck, and Goudsmit about spin; please read that.

Next is the story of Pauli's "neutron."

I told you in lecture 7 that around 1930 he wrote to several people about this idea. The letter which I quoted earlier was the one which was cited in Jensen's Nobel lecture, but Yasuo Hara of the Tokyo University of Education told me that there is a more detailed letter in a supplementary physics textbook for American high-school students. While the letter quoted in Jensen's lecture was perhaps aimed at theoreticians, this letter was for experimentalists, exhorting them to find such particles. Mr. Hara sent me a copy, so I quote it here.[3]

Zurich, December 4, 1930

Dear radioactive ladies and gentlemen,

I beg you to most favorably listen to the carrier of this letter. He will tell you that, in view of the "wrong" statistics of N and Li^6 nuclei and of the continuous beta spectrum, I have hit upon a desperate remedy to save the laws of conservation of energy and statistics. This is the possibility that electrically neutral particles exist which I will call neutrons, which exist in nuclei, which have a spin $1/2$ and obey the exclusion principle . . . The mass of the neutrons should be of the same magnitude as the electron mass . . . Thus, dear radioactive ones, examine and judge . . .

Your most obedient servant,

W. Pauli

According to this letter, we see that Pauli was trying to resolve the difficulty not only of the beta spectrum but also of the contradiction of the statistics and spin of the N nucleus (and the ^6Li nucleus) by introducing his "neutron." By the way, *radioactive lady* apparently refers to Lise Meitner.

Now I shall change the subject slightly. I heard from Professor Nishina the story that Pauli always said Dirac's way of thinking was acrobatic. The story about "Pauli's sanction" was also from Professor Nishina. Supposedly, when Heisenberg arrived at the new idea to use quantities whose products are non-commutative (it was soon found that they are matrices), he first sought Pauli's opinion. For this, Pauli is said to have given his sanction right away. Professors

3. For another translation of Pauli's letter to the Tübingen meeting, see Pais, *Inward Bound*, p. 315.

Nishina and Sugiura of RIKEN, on whom I shall remark later, very often talked to me about European scholars and their research lives.

Among Professor Nishina's stories there is naturally the account of his difficult but inspiring experience in deriving together with Klein the Klein-Nishina formula. When they did these derivations, they independently calculated to some point and then compared their results, and independently calculated again and compared, and repeated the process. Nowadays this method is standard, but at that time it was new to us. We adopted the same method when I worked with Shoichi Sakata, Hidehiko Tamaki, and Minoru Kobayashi, and we had a hard time because each of us made mistakes, but Professor Nishina told us that when he was working with Klein, they also had a hard time because of discrepancies in their calculations. When we learned it was not just we, we were a little bit relieved.

You might think that the Klein-Nishina formula had been derived using a method such as that described in Heitler's *The Quantum Theory of Radiation*, but that is not the case. Remember when they started the derivation, there had been no field quantization; they used the makeshift method of calculating the electron field and electromagnetic field classically and then applying the correspondence principle to the results to translate them into quantum theory. Note that treating the electron field classically automatically means that it is based on Schrödinger's idea that the electron wave is a wave in three-dimensional space. This idea of treating the interaction between the electron wave and the electromagnetic field using the correspondence principle was developed independently by Klein and Gordon around 1926; they first considered the relativistic electron wave equation for a scalar field and calculated, among other things, the Compton effect. For this reason the scalar equation for the electron has been named the Klein-Gordon equation, but what they aimed for was not the equation itself but the explanation of the Compton effect. Thus, it was quite natural that as soon as the Dirac equation appeared, Klein and Nishina wanted to calculate the Compton effect using the Dirac equation.

As I said earlier, by using the correspondence principle Klein and Gordon salvaged Schrödinger's unsuccessful idea to consider the electron wave as a wave in three-dimensional space rather than in configuration space. We can regard this method as the forerunner of the field quantization, which appeared soon after. Since the classical description of the motion of a particle and the application of the correspondence principle to the solution—the methods of the old quantum theory—were formulated by Heisenberg into matrix mechanics, it is naturally expected that the method of Klein and Gordon, which applies the correspondence principle to the classical theory of the wave field, will be formulated by considering the field quantities as q-numbers. In fact, it seems that Klein had had the idea to quantize ψ since about 1926, and he completed this work in 1927 together with Jordan (this was discussed in lecture 6), who had had the same idea, although the priority was taken by Dirac's acrobatics.

Anyhow, if we use the method in Heitler's book, the Klein-Nishina formula would have perhaps been derived in ten days. If we use Feynman's method, probably three hours suffices. However, when Professor Nishina did it with Klein, it apparently was a very lengthy calculation. I might add that soon after the Klein-Nishina paper there was a paper in which Professor Nishina as sole author discussed the polarization of the scattered γ-ray, but this calculation seems to have been even more complicated. Therefore C. Møller, who was then a graduate student, checked the calculation. Of course, Professor Nishina thanked Møller at the end of the paper.

The problem of whether Schrödinger's ψ is a wave in the configuration space or a wave in three-dimensional space also worried us when we were third-year undergraduate students at Kyoto University and were studying quantum mechanics. That was in 1928, and by then the Jordan-Klein paper had been already published, but to a novice just starting to understand quantum mechanics as a third-year student, this paper was not known (nor was there any professor who taught these things). I was just saying, "I don't understand, I don't understand," but apparently Yukawa believed ψ must be a wave in three-dimensional space, and he tried hard to solve the problem of He without using ψ in six-dimensional space. As a result of that, he arrived at the idea to set up a Schrödinger equation in three-dimensional space by considering the electric field due to the charge density $-e\psi^*\psi$ in addition to the nuclear electric field Ze/r and solved the equation. I remember he told me that he obtained He energies that agreed pretty well with experiments. Perhaps you see this already, but in essence this is the Hartree approximation.

In relation to these, I remember that towards the end of our third undergraduate year, there was a journal club in which the students introduced papers they had read.[4] We made a public report for the first time in our lives. I reported on Heisenberg's "Mehrkörperproblem und Resonanz in der Quantenmechanik" (Many-body problem and resonance in quantum mechanics), and Yukawa chose Klein's paper "Elektrodynamik und Wellenmechanik von Standpunkt des Korrespondenzprinzip" (Electrodynamics and wave mechanics from the standpoint of the correspondence principle). This is the aforementioned paper by Klein, and as I said, he regarded the electron wave as a wave in three-dimensional space, so it was quite natural that Yukawa chose this paper. On the other hand, as for me, the paper by Heisenberg that I chose discusses how ψ obeys Bose statistics or Fermi statistics, depending on whether ψ is symmetric or antisymmetric with respect to the interchange of particles, and ψ here is certainly a function in the configuration space. From this it seems that I was attracted to the idea of ψ being in the configuration space. However, I think I was attracted to this paper more because of Heisenberg's expert use of analogy. I was attracted by the deftness of the analogy in which he started from the

4. At that time, the baccalaureate program in a Japanese college lasted for three years.

resonance of two pendula, which is a very ordinary, everyday phenomenon, and gradually proceeded to the sophisticated problem of the symmetry of ψ and the statistics of the particle. The fact that I was not attracted at all to Dirac's paper, which discusses the same problem, attests to this.

A little bit later, I met with Yukawa in the library, where he opened on a table the issues of *Zeitschrift für Physik* in which the Jordan-Klein paper and the Jordan-Wigner paper were published and informed me that there was this surprising work that, if ψ in the three-dimensional space is quantized by the canonical commutation relation or anticommutation relation, then we could obtain exactly the same conclusion as when we took the symmetric or antisymmetric wave function using ψ in configuration space. Thus I also read those papers right away and found that the murky problem of three-dimensional wave versus multidimensional wave had been completely and elegantly answered.

In the year 1929, when we were graduated from the university, Heisenberg and Dirac came to Japan in September. They gave lectures in Tokyo and Kyoto. I got up my nerve, went to Tokyo, and listened to the lectures. These lectures were given from September 2 through 9 at the University of Tokyo and at RIKEN. Heisenberg's lectures were entitled (1) "Theory of Ferromagnetism"; (2) "Theory of Conduction" (Bloch's theory of electric conduction); (3) "Retarded Potential in the Quantum Theory" (the famous Heisenberg-Pauli theory); and (4) "The Indeterminacy-Relations and the Physical Principles of the Quantum Theory." The titles of Dirac's lectures were (1) "The Basis of Statistical Quantum Mechanics" (the density matrix); (2) "Quantum Mechanics of Many-Electron Systems" (representing the permutation operator of electron coordinates by spin variables and its applications); (3) "Relativistic Theory of the Electron" (needless to say, this is the story of the Dirac equation); and (4) "The Principle of Superposition and the Two-Dimensional Harmonic Oscillator." Lectures 1, 2, and 3 were given at the University of Tokyo, and lectures 4 were given at RIKEN. As you see, the content of the lectures was at the forefront of contemporary physics.

Miraculously, I remember, I could more or less understand the content of the lectures because fortunately I had already looked through papers related to these talks. (I shall tell you later, however, that this required great labor.) This was the first time I had come from rural Kyoto to Tokyo and seen in person distinguished people like Professor Hantaro Nagaoka, Professor Nishina, and Professor Sugiura and also the brilliant graduates of the University of Tokyo, who obviously looked very bright. I listened to the lectures, hiding myself toward the last row of the room, overwhelmed by those luminaries. There was one senior student who had finished at the Third High School of Kyoto and was graduated from the physics department of the University of Tokyo, and he pointed out to me that that was Professor Nishina and those were Masao Kotani

Figure 12.1 In September 1929, when I was graduated from the university, Heisenberg and Dirac visited Japan. From the left: Nishina, skipping two, Heisenberg, Nagaoka, and Dirac; Sugiura is on the far right.

and Tetsuro Inui, who were studying quantum mechanics in Professor Nishina's colloquium. He encouraged me to get acquainted with these people, but I simply stayed shy. In this atmosphere, I clearly remember the question Dirac asked Heisenberg after the third lecture. As you probably know, Heisenberg and Pauli in their theory introduce the condition $\nabla \cdot \mathbf{E} = 4\pi\rho$, not as a relation between the q-numbers, but as an additional condition for the state vector ψ, $(\nabla \cdot \mathbf{E})\psi = 4\pi\rho \cdot \psi$. Dirac asked the question whether the eigenvalue 0 of $\nabla \cdot \mathbf{E} - 4\pi\rho$ is discrete or continuous. Apparently this was an unexpected question for Heisenberg. He could not give an answer right away and after thinking for a while answered, "It is probably continuous."

I remember on the last day of the lectures at the University of Tokyo, Professor Nagaoka got up and raved about how Heisenberg and Dirac in their twenties had accomplished such a major thing as the establishment of a new theory and deplored that in Japan the physicists were still picking up the chaff and bran of Europe and America, and that students were just copying lectures, which was terrible. "You guys should emulate Heisenberg and Dirac." (Professor Nagaoka ranted this in his Nagaoka-English, and I could not quite hear it accurately, so this may be my arbitrary translation.)

I used to regret sometimes that I had chosen quantum theory as my major subject when I was a third-year undergraduate student and had to choose my major field. At that time, there was no textbook on quantum mechanics. There

Figure 12.2 One year after I was graduated from the university, Professor Nishina visited Kyoto. Front row, second from the right: Nishina, Kimura. Back row, second from the right: Tomonaga, Yukawa.

were only books like *Collected Papers of Schrödinger* or Born's *Problems in Atomic Physics*, and most studying was done by looking up the original papers one after the other. As I did that, I found that each paper quoted lots of others, and unless I read them, I could not understand what was written there. For this reason I drowned in a sea of many papers. Furthermore, I was in delicate health at that time, and although I had received a bachelor's degree, I was in a totally neurotic state. Thus I often thought of quitting quantum mechanics, but after a year and a half or so, I found that I had caught up to the level that I could more or less understand the lectures of Heisenberg and Dirac. However, by the time I caught up, the enemy had advanced. Professor Nagaoka's pep talk really did not get me anywhere.

It was quite natural that, even among the conservative professors at Kyoto University, who were completely out of touch, the atmosphere arose that something must be done in view of the fact that the new physics of the quantum theory was spreading throughout the world like wildfire. At that time in Kyoto

University there was a professor of spectroscopy Masamichi Kimura, whose name was known even outside Japan. He was an experimentalist, and apparently he initially did not like theoretical physics that much, but after visiting abroad and witnessing the status of physics in Europe and America, he thought that we should not be studying only classical physics in Japan. Luckily he was also a chief researcher at RIKEN; he asked Professor Sugiura of RIKEN to give an intensive series of lectures on quantum mechanics at Kyoto University. So Professor Sugiura came to Kyoto, I think around the beginning of 1930. I remember it was a cold time, and we listened to the lecture in a room heated by a stove. After that, I think in early summer of the next year, Professor Nishina came to give lectures.

Professor Sugiura learned that Yukawa and I were studying quantum mechanics. He said that if we liked, he would propose some topic of study. I told you that by the summer of 1929 I, at a very slow pace, had somehow caught up to the quantum mechanics, but it is altogether quite different to do some original work. The theoretical framework of quantum mechanics had been more or less completed, and the problems of the atom had been almost all solved, so there was not much left to do in those fields. Therefore, I got somewhat interested in molecules, studied the work of Hund, and looked for an interesting problem related to molecular structure. I found, however, that even there the physical problems were more or less done and all that was left was more suitable for chemists. Therefore, it seemed to me that the only areas where I could work were in the fields of solid state physics, nuclear physics, and relativistic quantum mechanics. After racking my brain for which direction to take, I came up empty-handed. I could not help realizing that whatever direction I took, my ability was not sufficient. Therefore, when Professor Sugiura proposed to give me a topic, I wanted to use this occasion to decide in which direction to proceed.

On the other hand, from early on Yukawa seemed to have determined to study either nuclear physics or relativistic quantum mechanics. (Mr. Yukawa, please correct me if I am wrong.) By self-study, he was attacking the theory of the hyperfine structure of spectra due to nuclear spin and other problems. However, he probably also wanted to hear what topic Professor Sugiura would give him, and we both went to Professor Sugiura's office.

The project the professor gave me was the problem of the Na_2 molecule (I do not know whether he knew that I was interested in molecules), which was the task of applying Heitler and London's theory of H_2 to Na_2. I was already lukewarm about molecules at that time, but I thought it would be instructive to do something by myself rather than reading other people's papers, and therefore, I dared to say, "Let me do it." This was a numerical calculation from the beginning, and it did not look that instructive. (It was good training for perseverance, however.) Moreover, it was like picking up the crumbs of Professor Sugiura's work, and it was not inspiring at all. Also, when the calculations were carried out, there appeared many awkward results, and the whole thing did not work.

Professor Sugiura also gave a topic to Yukawa to study, namely, to theoretically explain the peculiar result of Bergen Davis' experiment. However, if we examine the experiment in detail, it is dubious. It was very likely a botched experiment. After we had been stuck in this situation for a while, Professor Nishina came to lecture at Kyoto University.

Heisenberg's book *Physikalische Prinzipien der Quantentheorie* (The physical principles of quantum theory) was used as the text for Professor Nishina's lectures. When Professor Sugiura would lecture, he would write a long, long formula (called a confluent hypergeometric function) from one end to the other of a long blackboard, and he would lecture about his own work. It might have been a creative work, but it was too detailed for a beginner to make sense out of it. On the other hand, Professor Nishina's lecture, albeit much of it was from the book used and much credit should go to Heisenberg, was impressive, especially the discussions after the lectures. Professor Kimura's comment based on the idea of experimentalists, Professor Nishina's response, and so on provided an atmosphere in which even I, who was timid and could not talk on such an occasion so far, managed to ask Professor Nishina a question after a lot of hesitation. Professor Nishina was extremely kind in answering it and in listening to a novice such as I.

One day during his stay at Kyoto, Professor Nishina invited Yukawa and me to dinner. When we talked about the work given by Professor Sugiura, Nishina said that Bergen Davis' experiment had been proved wrong and that in experiments in which scintillation was used, the experimental results were often affected by the mind-set of the experimenters.

As for my work, well . . . , the professor said that there were many other interesting things. Nishina showed us a recent letter he had received from Klein, and he talked about Bohr's statement in the letter that quantum mechanics was not valid in the atomic nucleus. According to Bohr's idea, quantum mechanics was more advanced than classical mechanics in that the effect of the observation on the object could not be smaller than h, but since the observing apparatus itself was composed of protons and electrons, there must be some additional intrinsic limitation on the operation of observation. Because of such limitation, the present quantum mechanics, which did not take that into account, would not be valid in the nucleus. Nishina told us that as a result of this limitation, the electron must lose its identity in the nucleus—that was Bohr's idea. (I write these as if I clearly heard these words, but it is hardly likely that at that time I could clearly comprehend such a difficult subject. So I must be unconsciously touching it up just like a person who talks about a dream after waking up.)

After all these events, a letter from Professor Nishina arrived at the beginning of 1932. In this letter he asked whether I wanted to study in his lab at RIKEN.

I was very hesitant, and I had reservations, not being sure whether I could live up to the expectations of those world-famous professors at a top-notch place like RIKEN, but I also hoped that this would be a good chance and I should jump at the opportunity. I wrote these feelings honestly to Professor Nishina, to which he replied, "Why don't you come for two or three months to try it. You can continue that calculation on Na$_2$, or there are many other interesting projects, so you can come and decide." Thus I went to Tokyo at the end of April 1932 and became a member of Nishina's lab in RIKEN.

When I arrived at Nishina's lab, he asked me what was going on with Na$_2$. When I said that it was not getting anywhere, he told me that there was one calculation he would like me to do on the neutron, and he explained in detail what this particle called neutron was. I went to Professor Sugiura and said that I was now working in Nishina's lab and that I was doing the calculation on neutrons. Sugiura congratulated me and said that would be much more interesting than Na$_2$, and he told many stories from Europe. He talked about how his work on confluent hypergeometric functions was appreciated by Pauli and described the atmosphere of Göttingen University where he had worked. (At that time Dirac and Oppenheimer were there.) I think it was at this time that I heard the story of Kronig and Pauli.

The problem which Professor Nishina proposed was to calculate the cross-section for excitation and ionization when neutrons pass through some material. In 1932 the neutron was discovered, but the true nature of cosmic rays was not yet understood. Professor Nishina had the idea that they were neutrons and asked me to do such a calculation. At that time he had not yet received Heisenberg's paper on nuclear structure, and therefore he still did not know about the nuclear force. Nishina thought that although the neutron is neutral, it must have electric or magnetic moments, and such a dipolar field might interact with electrons and thus excite or ionize atoms.

In order to do this calculation, we need the wave equation of a particle which is neutral but which has an electric or magnetic moment. At that time *Handbuch der Physik*, which I mentioned in the last lecture, was not yet published, and I did not know of the Pauli term. However from the relativistic requirements, it must have the form $M' \rho_2[(\boldsymbol{\sigma} \cdot \mathbf{H}) - i(\boldsymbol{\alpha} \cdot \mathbf{H})]$, and therefore I used the Dirac equation for a neutral ($e = 0$) particle with what are now called Pauli terms. (This was not a big deal; I am just telling you what I did.) In the meantime, Heisenberg's paper arrived in Tokyo. It was found that the interaction between a neutron and a nucleus is much larger than that between a neutron and an electron. Furthermore, when deuterium was discovered, it was immediately obvious that solving this two-body problem and determining the properties of the nuclear force were much more important, so we switched our project.

For these reasons I started calculations related to phenomena such as the binding energy of the deuteron and the scattering or capture of a neutron by a proton. Heisenberg regarded the nuclear force as an exchange force and introduced the potential $J(r)$ for it. He figured that the nuclear force must act only over a very short range and that the potential goes to zero if $r \geq 10^{-13}$ cm. Now, it was necessary to determine the magnitude of the potential. Since the binding energy of the deuteron was known experimentally, it was possible to determine the magnitude of the nuclear force so that the theoretical value for the binding energy agreed with experiment. Using the potential, we can discuss neutron scattering and capture. As more and more experimental facts surfaced, both the cross-section of elastic collision and that of the capture of a neutron by a proton were found to be abnormally large for slow neutrons, and this drew Professor Nishina's attention.

However, when I took the potential $J(r)$ determined by the method mentioned earlier from the binding energy of the deuteron and used it for scattering, this conclusion did not emerge. I assumed a variety of forms for $J(r)$, such as the potential wells $-e^{-r/a}$, $-e^{-r/a}/r$, etc., but the conclusion was almost independent of the assumed form. Then it occurred to me that if in the two-body system of a proton and a neutron, in addition to the S level of deuterium, there should exist another S state whose energy was nearly zero, then the S wave in the incident neutron would resonate with it, and the zero-energy neutron would show a very large scattering cross-section. Moreover, I found out that such a level did not result from Heisenberg's exchange force alone, but if we added the exchange force proposed by E. Majorana (I touched upon the Majorana force in lecture 10), such a level was possible. Therefore, I wanted to determine the ratio of the Heisenberg force and the Majorana force from this scattering experiment. When I did this, the idea worked very well as far as elastic scattering was concerned, and I succeeded in determining the ratio of the two forces.

I was quite elated with this achievement, and Professor Nishina was also satisfied, and our results were reported successively from the annual meeting of the Physico-Mathematical Society of Japan in Sendai, 1933, to the autumn meeting of RIKEN in 1935. However, probably because he was so busy with a variety of experimental work, Professor Nishina put off publishing this paper. While I was agonizing about this, Bethe and Peierls did exactly the same thing and published it. I was extremely upset,[5] and I was livid with Professor Nishina.

While elastic scattering was explained by this idea, neutron capture was not. In my calculation the cross-section for capture became zero if the energy of the neutron was zero. The reason for this is that for capture $h\nu$ must be emitted, but according to the normal selection rules, such a process must be $P \rightarrow S$, and therefore even if there is another S energy level very close to zero, there is no influence on the P wave. As a result of this, if the energy of the incident neutron is zero, the cross-section also becomes zero. The story would be different if there existed a P level very close to zero energy, but then we would have to change $J(r)$ drastically in order to overcome the centrifugal

5. Literally, I felt as if I were biting my navel.

force, and if we do this, all the other agreements of the theory with experiment would be lost.

However, there was another possibility: since the neutron and proton have magnetic moments, it is possible to emit $h\nu$ via a magnetic dipole in addition to the usual emission following the selection rule (namely, emission of $h\nu$ via an electric dipole), and for this emission $S \to S$ is allowed. Fermi was the first to point this out. When I saw Fermi's paper, I was devastated. We had been prisoners of the ordinary common sense of spectroscopy and were thus blinded by the conventional notion that emission of $h\nu$ due to a magnetic dipole was very small and approximately forbidden. Our only consolation was that Bethe and Peierls did not think about this possibility either; it is quite difficult to be free of preconceptions.

There are many other stories about my days in Professor Nishina's laboratory. For example, calculations were made for a variety of phenomena related to the positron using the hole theory. However, if I talk too much, we shall run out of time, so I shall limit myself to stories related to the subjects discussed in this book. However, I can say with certainty that this period while I was Professor Nishina's assistant was decisive in determining the direction of my research, i.e., to proceed in the theory of atomic nuclei, cosmic rays, and quantum electrodynamics. Before that, I had been astray in Kyoto.

While we were doing this type of work in Tokyo from around 1933 to 1935, Yukawa in Osaka was incubating his idea about the meson. Apparently, he wanted to create the theory of β-decay soon after Heisenberg's paper on the exchange force was published in 1932. Namely, Yukawa's idea was to describe the disintegration using Heisenberg's isospin according to Heisenberg's idea that a neutron changes into a proton by emitting an electron. I do not remember precisely when, but I vaguely remember that Yukawa sent a write-up of his idea to Professor Nishina, who shared it with me. If I remember correctly, he quantized the electron field as a Fermi field, without the idea of the neutrino, and as a result, I had an impression that Yukawa was having considerable difficulty because of the inconsistencies which pop up here and there in the theory. In the meantime Fermi proposed his theory of β-decay. Therefore, many people tried to use Fermi's theory to explain the force between a proton and a neutron as due to the exchange of an electron-neutrino pair, but they found that it did not work.

I think it was at the Physico-Mathematical Society's annual meeting in Sendai in 1933 that Yukawa told me his idea of the particle with a mass one hundred times that of the electron by drawing a formula with a stick on the ground of the athletic playing field. I think he said that he could explain the nuclear force if we could assume such a weird particle existed. I think I also heard from somebody at Osaka University that he was very much concerned about what we were reporting in the conference because the title of the talk by Professor

Nishina and me happened to be "A Note on the Force between the Proton and the Neutron." By the way, the title of Yukawa's paper was "On the Force between Elementary Particles." Just from the title, our work also sounded like the theory of mesons. My recollection may not be accurate, but I hope it entices Yukawa to give his version of the facts.

I told you earlier in lecture 10 that the theory of the meson was also developed in Europe around 1936–1937, and in 1939 it was decided that a new Solvay Congress was to be held with the development of this new theory as its main theme. Yukawa was the first Japanese invited to this conference, and he came to Europe. Shortly before the meeting he visited me in Leipzig, where I was studying at that time. Unfortunately for Yukawa, it was summer recess at the university, and neither Heisenberg, Hund, nor other young colleagues were there. Therefore, he just looked through the newly published journals in the physics department library and returned to Berlin. Meanwhile, the Second World War started in Europe. We received notice from the embassy in Berlin advising all Japanese in Germany to repatriate, and thus both Yukawa and I were fated to leave Europe. Needless to say, the Solvay Congress was canceled.

In this Solvay Congress Pauli was to have given a talk on the grand theme of the quantization of a generally applicable relativistic field, in which he was to have laid out the general aspects of the theory including the relation of spin and statistics, which I discussed in lecture 8, and other magnificent discussions in his favorite, intricate way. That of course included the commutation relations in the four-dimensional space. A manuscript of the Solvay talk prepared by Pauli was sent to Yukawa, and after he returned to Japan, Yukawa copied it for distribution in the country. When I established my super-many-time theory during the war, one of the important points of the theory was the four-dimensional commutation relations, and had I not had this manuscript, I would have had to work considerably more to create my theory. Indeed, the four-dimensional commutation relations for the electromagnetic field were already available in Pauli and Jordan's 1927 paper. The four-dimensional commutation relations for an arbitrary Bose field and an arbitrary Fermi field were not discussed in detail in any papers except in Pauli's manuscript for the Solvay Congress, and if I had not had it, it would have taken me much longer to create the super-many-time theory.

Now I am going to conclude this lecture. Thank you very much for your attention for such a long time. I also wish to thank Professor Nitta and others who supplied a variety of interesting materials. My anecdotes related to Professor Yukawa are based on my somewhat uncertain memory, and perhaps quite a bit of my imagination and subjectivity may have crept into it, which I am afraid might annoy Yukawa a little bit. However, as I told you earlier, it is highly desirable that how the great masterpiece of the prediction of the meson's existence took shape from the murk be described by its creator, and I hope this talk of mine will invite him to do so. I simply ask the indulgence of Master Yukawa on this matter.

Finally, Mr. Ishikawa of *Shizen* has generously accepted the painstaking work of tracking down and copying the papers which I have requested. Not only that, he often discovered papers and other literature which were unknown to me and informed me of their existence—it was just like Yukawa informed me about the papers of Jordan and Klein and of Jordan and Wigner. (Mr. Ishikawa, don't be shy!) This literature was of great help to me.

EPILOGUE

The lectures *Spin wa meguru* were initially published in the January through October 1973 issues of *Shizen*, and I expanded and revised them quite a bit to make them into this book. I started out this series intending to describe the subject concisely and simply, but as I kept on writing, the manuscript grew, and the contents became terribly detailed. As I was reading many of the old papers, I often remembered the old days when I first read them, and I could not resist the temptation to relive how I felt, what I thought, what I found very difficult at that time. Thus my pen moved by itself, on and on. Even those subjects which could have been treated lightly started to involve many formulas which might burden readers, and the whole thing became rather unwieldy.

If I look up the word *spin* in an English dictionary, in addition to the meaning to rotate, it can mean twisting a fiber to form thread, and perhaps from there it can also mean to draw out something. Especially there is the idiom *spinning a yarn*, and this seems to describe an old sailor talking on and on about his youthful adventures. (Mr. Ishikawa[1] informed me of this one.)

Anyhow, the period from the 1920s through the 1940s, which is the stage of this book, is a particularly rich period in the history of physics, especially worth recording, when quantum mechanics gradually matured. The special characteristic of this period was the youth of its major players. For example, Pauli was twenty-five years old when he published his exclusion principle, and Heisenberg was twenty-four years old when he arrived at the idea of matrix mechanics. Dirac was twenty-six years old when he discovered his equation. (I have added the year of birth under the photos of people in this book so you can also check the ages of others when referring to the bibliography and publication dates therein.) Several years later this youthful trend made its way

---

1. An editor of *Shizen*. He helped Tomonaga to prepare this book.

to Japan. I therefore thought it worthwhile that this period be recorded by somebody who passed his scientifically formative years during that time and directly experienced part of the history. I therefore have augmented my articles published in *Shizen* for more completeness and publish them as a volume in this *Shizen Selected Series*.

After finishing this work I started to worry that somebody might ask, To which readers and for what purpose did you write this book? This book has grown while I indulged my tastes without worrying too much about the readers.

May 1974
Sin-itiro Tomonaga

SHORT BIOGRAPHY OF SIN-ITIRO TOMONAGA

1906 Born March 31, 1906, in Tokyo, as the first son of Sanjuro and Hide Tomonaga. Father was a professor of philosophy at Shinshu University in Tokyo.

1907 Father was appointed to Kyoto Imperial University, and the family moved to Kyoto.

1926–29 Undergraduate student in Kyoto Imperial University. Majored in physics. Stayed at the university as an unpaid junior assistant.

1932 Joined Nishina's laboratory at RIKEN (Institute of Physical and Chemical Research) as a research associate. Started research on cosmic rays and nuclear physics.

1937–39 Studied nuclear physics under Heisenberg at Leipzig University. Became an assistant at RIKEN. Received Doctor of Science from the University of Tokyo for the study of the nucleus.

1940 Became an instructor at Tokyo Bunrika University (later Tokyo University of Education, now the University of Tsukuba). Married to Ryoko.

1941 Professor at Tokyo Bunrika University.

1943 Published super-many-time theory.

1947 Awarded the Asahi Prize for the development of the meson theory and the super-many-time theory.

1948 Received the Japan Academy Award for his study on the oscillation mechanism of magnetron with M. Kotani.

1949–50 Invited by Oppenheimer to the Institute for Advanced Study at Princeton. Studied many-body theory.

1952 Awarded the Order of Culture from Emperor

1965–62 President of Tokyo University of Education. "Became an administrator at last."

235

1963	Elected president of the Science Council of Japan.
1965	Awarded Nobel Prize in Physics with J. Schwinger and R. Feynman for fundamental studies in the field of quantum electrodynamics.
1969	Retired from Tokyo University of Education.
1974	*Spin wa meguru* published.
1978	Diagnosed as having cancer of the esophagus.
1979	Passed away on July 8.

The following books by Tomonaga are available in English:

1962	*Quantum Mechanics: Volume 1. Old Quantum Theory* trans M Koshiba (New York: Interscience).
1966	*Quantum Mechanics: Volume 2. New Quantum Theory* trans M Koshiba (New York: Wiley).
1971	*Scientific Papers of Tomonaga Volume 1* ed T Miyazima (Tokyo: Misuzu Shobo).
1976	*Scientific Papers of Tomonaga Volume 2* ed T Miyazima (Tokyo: Misuzu Shobo).

ANNOTATED BIBLIOGRAPHY

In order to make clear which references have been used for this story, I append this bibliography. Those papers which are quoted in my talk but are not in this list are either hearsay or indirectly quoted.

[*Translator's note:*

AS = *Atomic Spectra* ed W R Hindmarsh (New York: Pergamon, 1967); several reprints and several English translations of original papers.

CPWM = *Collected Papers on Wave Mechanics* E Schrödinger (New York: Chelsea 1982); English translations of papers.

EQED = *Early Quantum Electrodynamics: A Sourcebook* ed A I Miller (Cambridge: Cambridge University Press, 1995); several reprints and several English translations of original papers.

SQM = *Sources of Quantum Mechanics* ed B L van der Waerden (Amsterdam: North-Holland 1967; repr. New York: Dover 1968); several reprints and several English translations of original papers.

SPFNP = *Selected Papers in Foundations of Nuclear Physics* ed R T Beyer (New York: Dover, 1949); reprints of original papers; exhaustive bibliography.

SPQED = *Selected Papers on Quantum Electrodynamics* ed J Schwinger (New York: Dover 1958); reprints of original papers.]

Lecture 1
General summary of the atomic core, radiant electron, and atomic spectra
Hund F 1927 *Linienspektren und Periodisches System der Elemente* (Berlin: Springer), chap. 2.

Space quantization
Sommerfeld A 1916 Zur Theorie des Zeeman-Effekts der Wasserstofflinien, mit einem Anhang über den Stark-Effekt *Phys. Z.* **17** 491–507.

Internal quantum number, anomalous Zeeman effect, **Ersatzmodell**
Sommerfeld A 1920 Allgemeine spektroskopische Gesetze, insbesondere ein magnetooptischer Zerlegungssatz *Ann. Phys., Lpz.* **63** 221–263 [Some General Laws of Spectroscopy, and in particular a Theorem of Magneto-Optic Resolution *AS* excerpts 145–149]; 1923 Über die Deutung verwickelter Spektren (Mangen, Chrom usw.) nach der Methode der inneren Quantenzahlen *Ann. Phys., Lpz.* **70** 32–62; 1923 Spektroskopische Magnetonenzahlen *Phys. Z.* **24** 360–364.
Landé A 1923 Termstruktur und Zeemaneffekt der Multipletts *Z. Phys.* **15** 189–205 [Term Structure and Zeeman Effects in Multiplets *AS* 186–205]; 1923 Termstruktur und Zeemaneffekt der Multipletts: Zweite Mitteilung *Z. Phys.* **19** 112–123; 1923 Das Versagen der Mechanik in der Quantentheorie *Naturwissenschaften* **11** 725–726.
Pauli W 1923 Über die Gesetzmäßigkeiten des anomalen Zeemaneffektes *Z. Phys.* **16** 155–164; 1924 Zur Frage der Zuordnung der Komplexstrukturterme in starken und in schwachen äußeren Feldern *Z. Phys.* **20** 371–387.
Sommerfeld A 1922 Quantentheoretische Umdeutung der Voigtschen Theorie des anomalen Zeemaneffektes vom *D*-Linientypus *Z. Phys.* **8** 257–272.
Heisenberg W 1922 Zur Quantentheorie der Linienstruktur und der anomalen Zeemaneffekte *Z. Phys.* **8** 273–297.

Lecture 2
Fine structure formula
Sommerfeld A 1916 Zur Quantentheorie der Spektrallinien *Ann. Phys., Lpz.* **51** 125–167.

Level spacing of doublet terms
Heisenberg W 1922 Zur Quantentheorie der Linienstruktur und der anomalen Zeemaneffekte *Z. Phys.* **8** 273–297.
Landé A 1923 Zur Theorie der Röntgenspektren," *Z. Phys.* **16** 391–396; 1924 Das Wesen der relativistischen Röntgendubletts *Z. Phys.* **24** 88–97; 1924 Die absoluten Intervalle der optischen Dubletts und Tripletts *Z. Phys.* **25** 46–57.

Classically indescribable two-valuedness, exclusion principle
Pauli W 1925 Über den Einfluß der Geschwindigkeitsabhängigkeit der Elektronenmasse auf den Zeemaneffekt *Z. Phys.* **31** 373–385; 1925 Über den Zusammenhang des Abschlusses der Elektronengruppen im Atom mit der Komplexstruktur der Spektren *Z. Phys.* **31** 765–783; 1964 Exclusion Principle and Quantum Mechanics *Nobel Lectures: Physics 1942–1962* (Amsterdam: Elsevier) 27–43.

Nonmechanical constraint
Bohr N 1923 Linienspektren und Atombau *Ann. Phys., Lpz.* **71** 228–288 (esp. p 276) [1977 Line Spectra and Atomic Structure *Niels Bohr Collected Works: Vol. 4 The Periodic System (1920–1923)* ed J Rud Nielsen (Amsterdam: North-Holland) 611–656].

Self-rotating electron
Uhlenbeck G E and Goudsmit S A 1925 Ersetzung der Hypothese vom unmechanischen Zwang durch eine Forderung bezüglich des inneren Verhaltens jedes einzelnen Elektrons *Naturwissenschaften* **13** 953–954; 1926 Spinning Electrons and the Structure of Spectra *Nature* **117** 264–265 [*AS* 253–258].

Kronig R d L 1926 Spinning Electrons and the Structure of Spectra *Nature* **117** 550; 1960 The Turning Point *Theoretical Physics in the Twentieth Century: A Memorial Volume to Wolfgang Pauli* ed M Fierz and V F Weisskopf (New York: Interscience) 5–59.

Goudsmit S A 1972 Guess Work: The Discovery of the Electron Spin *Delta* **15** 77–91.

Thomas factor

Thomas L H 1926 The Motion of the Spinning Electron *Nature* **117** 514; 1927 On the Kinematics of an Electron with an Axis *Philosoph. Mag.* **3** 1–22.

Ersatzmodell *based on the hypothesis of self-rotating electron*

Hund F 1927 *Linienspektren und Periodisches System der Elemente* (Berlin: Springer) chap. 3.

Relations among Kronig, Uhlenbeck and Goudsmit, and Pauli

van der Waerden B L 1960 Exclusion Principle and Spin *Theoretical Physics in the Twentieth Century: A Memorial Volume to Wolfgang Pauli* ed M Fierz and V F Weisskopf (New York: Interscience) 199–244.

Lecture 3
Equivalence of wave mechanics and matrix mechanics

Schrödinger E 1926 Über das Verhältnis der Heisenberg-Born-Jordanschen Quantenmechanik zu der meinen *Ann. Phys., Lpz.* **79** 734–756 [On the Relation between the Quantum Mechanics of Heisenberg, Born, and Jordan, and mine *CPWM* 45–61].

Transformation theory of quantum mechanics

Dirac P A M 1927 The Physical Interpretation of Quantum Dynamics *Proc. R. Soc.* A **113** 621–641.

de Broglie—Einstein relation

de Broglie L 1925 Recherches sur la Théorie des Quanta *Ann. Phys., Paris* **3** 22–128.

Klein-Gordon equation

Schrödinger E 1926 Quantisierung als Eigenwertproblem *Ann. Phys., Lpz.* **81** 109–139 (esp. § 6) [Quantisation as a Problem of Proper Values (Part IV) *CPWM* 102–123].

Gordon W 1926 Der Comptoneffekt nach der Schrödingerschen Theorie *Z. Phys.* **40** 117–133.

Klein O 1927 Elektrodynamik und Wellenmechanik vom Standpunkt des Korrespondenzprinzips *Z. Phys.* **41** 407–442.

Matrix mechanics of spin

Heisenberg W and Jordan P 1926 Anwendung der Quantenmechanik auf das Problem der anomalen Zeemaneffekte *Z. Phys.* **37** 263–277.

Pauli equation for spin
Pauli W 1927 Zur Quantenmechanik des magnetischen Elektrons Z. *Phys.* **43** 601–623.

Dirac equation
Dirac P A M 1928 The Quantum Theory of the Electron *Proc. R. Soc.* A **117** 610–624.

Derivation of the fine structure formula by the Dirac equation
Darwin C G 1928 The Wave Equations of the Electron *Proc. R. Soc.* A **118** 654–680.
Gordon W 1928 Die Energieniveaus des Wasserstoffatoms nach der Diracschen Quantentheorie des Elektrons Z. *Phys.* **48** 11–14.

Why Dirac did not derive the fine structure formula
Dirac P A M 1971 *The Development of Quantum Theory: J. Robert Oppenheimer Memorial Prize Acceptance Speech* (New York: Gordon and Breach).

Lecture 4
Symmetry of the wave function and statistics of particles

Heisenberg W 1926 Mehrkörperproblem und Resonanz in der Quantenmechanik Z. *Phys.* **38** 411–426; 1927 Mehrkörperproblem und Resonanz in der Quantenmechanik. II **41** 239–267.
Dirac P A M 1926 On the Theory of Quantum Mechanics *Proc. R. Soc.* A **112** 661–677.

Theory of band spectra; specific heat of molecules
Hund F 1927 Zur Deutung der Molekelspektren. I Z. *Phys.* **40** 742–764; 1927 Zur Deutung der Molekelspektren. II **42** 93–120.
Kronig R d L 1930 *Bandspectra and Molecular Structure* (Cambridge: Cambridge University Press).

Existence of nuclear spin
Pauli W 1924 Zur Frage der theoretischen Deutung der Satelliten einiger Spektrallinien und ihrer Beeinflussung durch magnetische Felder *Naturwissenschaften* **12** 741–743; 1964 Exclusion Principle and Quantum Mechanics *Nobel Lectures: Physics 1942–1962* (Amsterdam: Elsevier) 27–43.
Goudsmit S A 1961 Pauli and Nuclear Spin *Physics Today* **14** (June) 18–21.

Experiment on H₂ band spectrum
Hori T 1927 Über die Analyse des Wasserstoffbandenspektrums im äußersten Ultraviolett Z. *Phys.* **44** 834–854.

Specific heat of H₂, proton statistics and spin
Dennison D M 1927 A Note on the Specific Heat of the Hydrogen Molecule *Proc. R. Soc.* A **115** 483–486; [1974 Recollections of Physics and Physicists during the 1920's *Am. J. Phys.* **42** 1051–1056].

Lecture 5
Old interpretation of alkaline earth spectra, especially the necessity of strong spin-spin interaction

Hund F 1927 *Linienspektren und Periodisches System der Elemente* (Berlin: Springer) chap. 4 (esp. § 21).

New interpretation of alkaline earth spectra; apparent spin-spin interaction

Heisenberg W 1926 Über die Spektra von Atomsystemen mit zwei Elektronen *Z. Phys.* **39** 499–518 (esp. II. § 1) [The Spectra of Atomic Systems with Two Electrons *AS* 219–242].

Quantum theory of ferromagnetism

Heisenberg W 1928 Zur Theorie des Ferromagnetismus *Z. Phys.* **49** 619–636.

Einstein–de Haas experiment

Einstein A and de Haas W J 1915 Experimenteller Nachweis der Ampèreschen Molekularströme *Verh. Deutsch. Phys. Ges.* **17** 152–170.

Barnet S J 1935 Gyromagnetic and Electron-Inertia Effects *Rev. Mod. Phys.* **7** 129–166.

Particle exchange as a physical quantity

Dirac P A M 1929 Quantum Mechanics of Many-Electron Systems *Proc. R. Soc.* A **123** 714–733; 1963 *Principles of Quantum Mechanics, 4th Ed* (Oxford: Oxford University Press), chap. 9.

Lecture 6
Second quantization

Dirac P A M 1927 The Quantum Theory of the Emission and Absorption of Radiation *Proc. R. Soc.* A **114** 243–265 [*SPQED*, 1–24].

Jordan P and Klein O 1927 Zum Mehrkörperproblem der Quantentheorie *Z. Phys.* **45** 751–765.

Jordan P and Wigner E P 1928 Über das Paulische Äquivalenzverbot *Z. Phys.* **47** 631–651 [*SPQED*, 41–61].

Quantization of the electromagnetic field

Born M, Heisenberg W, and Jordan P 1926 Zur Quantenmechanik. II *Z. Phys.* **35** 557–615, (esp. Chap. 4, § 3) [On Quantum Mechanics. II *SQM* 555–615].

Quantization of the Klein-Gordon field and the electromagnetic field

Pauli W and Weisskopf V 1934 Über die Quantisierung der skalaren relativistischen Wellengleichung *Helv. Phys. Acta* **7** 709–731 [The Quantization of the Scalar Relativistic Wave Equation *EQED* 188–205].

Quantization of the Dirac field and the electromagnetic field

Heisenberg W and Pauli W 1929 Zur Quantendynamik der Wellenfelder *Z. Phys.* **56** 1–61; 1930 Zur Quantendynamik der Wellenfelder. II *Z. Phys.* **59** 168–190.

Hole theory
Dirac P A M 1930 A Theory of Electrons and Protons *Proc. R. Soc.* A **126** 360–365.
Oppenheimer J R 1930 On the Theory of Electrons and Protons *Phys. Rev.* **35** 562–563.
Weyl H 1931 *Gruppentheorie und Quantenmechanik, 2 Aufl.* (Leipzig: S. Hirzel), chap. 4, § 13 [1931 *The Theory of Groups and Quantum Mechanics* trans H P Robertson (London: Meuthen; 1950 repr. New York: Dover)].

Many-time theory
Dirac P A M 1932 Relativistic Quantum Mechanics *Proc. R. Soc.* A **136** 453–464.

Lecture 7
Failure in an attempt to put the Dirac theory in tensorial form
Darwin C G 1928 The Wave Equations of the Electron *Proc. R. Soc.* A **118** 654–680.

Covariance of the two-component quantity with respect to rotation of spatial coordinate axes
Pauli W 1927 Zur Quantenmechanik des magnetischen Elektrons *Z. Phys.* **43** 601–623.

Covariance of the four-component quantity with respect to Lorentz transformation
Dirac P A M 1928 The Quantum Theory of the Electron *Proc. R. Soc.* A **117** 610–624.

Two-valued representation of the rotation group
Wigner E P 1959 *Group Theory and Atomic Spectra* (New York: Academic), chap. 15.
Weyl H 1931 *Gruppentheorie und Quantenmechanik, 2 Aufl.* (Leipzig: S. Hirzel), esp. § 16, III [1931 *The Theory of Groups and Quantum Mechanics* trans H P Robertson (London: Meuthen; 1950 repr. New York: Dover)].

Spinor algebra
van der Waerden B L 1932 *Gruppentheoretische Methode in der Quantenmechanik* (Berlin: Springer) [1974 *Group Theory and Quantum Mechanics* (Berlin: Springer)].
Laporte O and Uhlenbeck G E 1931 Application of Spinor Analysis to the Maxwell and Dirac Equations *Phys. Rev.* **37** 1380–1397.
Umezawa H 1956 *Quantum Field Theory* (Amsterdam: North-Holland) chap. 3, § 8.

The mysterious tribe
Ehrenfest P 1932 Einige die Quantenmechanik betreffende Erkundigungsfragen *Z. Phys.* **78** 555–559.

Lecture 8
Electric current vector and energy-momentum tensor for the Klein-Gordon field
Schrödinger E 1926 Quantisierung als Eigenwertproblem *Ann. Phys., Lpz.* **81** 109–139 (esp. § 6) [Quantisation as a Problem of Proper Values, part 4 *CPWM* 102–123].

Gordon W 1926 Der Comptoneffekt nach der Schrödingerschen Theorie *Z. Phys.* **40** 117–133.
Klein O 1927 Elektrodynamik und Wellenmechanik vom Standpunkt der Korrespondenzprinzips *Z. Phys.* **41** 407–442.

Electric current vector and energy-momentum tensor for the Dirac field

Gordon W 1928 Der Strom der Diracschen Elektronentheorie *Z. Phys.* **50** 630–632.

Spin and statistics of elementary particles

Pauli W 1940 The Connection between Spin and Statistics *Phys. Rev.* **58** 716–722 [*SPQED*, 372–378].

Four-dimensional commutation relations for the electromagnetic field

Jordan P and Pauli W 1928 Zur Quantenelektrodynamik ladungsfreier Felder *Z. Phys.* **47** 151–173.

Lecture 9
Discovery of the neutron and of deuterium

Brickwedde F G 1982 Harold Urey and the Discovery of Deuterium *Physics Today* **35** (September) 34–39.
Chadwick J 1965 The Neutron and Its Properties *Nobel Lectures: Physics 1922–1941* (Amsterdam: Elsevier) 339–348.

Band spectrum of deuterium and the statistics and spin of the deuteron

Murphy G M and Johnston H 1934 The Nuclear Spin of Deuterium *Phys. Rev.* **46** 95–98.

Proton magnetic moment

Estermann I and Stern O 1933 Über die magnetische Ablenkung von Wasserstoffmolekülen und das magnetische Moment des Protons. I *Z. Phys.* **85** 17–24.
Rabi I I, Kellog J M B, and Zacharias J R 1934 The Magnetic Moment of the Proton *Phys. Rev.* **46** 157–163.

Magnetic moment of the deuteron

Rabi I I, Kellog J M B, and Zacharias J R 1934 The Magnetic Moment of the Deuton *Phys. Rev.* **46** 163–165.

Ehrenfest-Oppenheimer rule

Ehrenfest P and Oppenheimer J R 1931 Note on the Statistics of Nuclei *Phys. Rev.* **37** 333–338.

Pauli's letter on the neutrino

Jensen J H D 1972 Glimpses at the History of the Nuclear Structure Theory *Nobel Lectures: Physics 1963–1970* (Amsterdam: Elsevier) 40–50.

Lecture 10
Nuclear structure, exchange force between nucleons, and isospin
Heisenberg W 1932 Über den Bau der Atomkerne. I Z. Phys. **77** 1–11; 1932 Über den Bau der Atomkerne. II Z. Phys. **78** 156–164; 1932 Über den Bau der Atomkerne. III Z. Phys. **80** 587–596.
Majorana E 1933 Über der Kerntheorie Z. Phys. **82** 137–145.

Nuclear force between proton and proton
Tuve M A, Heydenberg N, and Hafstad L R 1936 The Scattering of Protons by Protons Phys. Rev. **50** 806–825.
Breit G, Condon E U, and Present R D 1936 Theory of Scattering of Protons by Protons Phys. Rev. **50** 825–845.

Nuclear energy levels of ^{14}C, ^{14}N, ^{14}O and the isotropy of the isospin space
Brink D M 1965 Nuclear Forces (Oxford: Pergamon Press) § 46.

Theory of β decay
Fermi E 1934 Versuch einer Theorie der β-Strahlen. I Z. Phys. **88** 161–177 [SPFNP 45–61; trans Wilson F D 1968 Fermi's Theory of β-Decay Am. J. Phys. **36** 1150–1160].

Meson theory of nuclear force
Yukawa H 1935 On the Interaction of Elementary Particles. I Proc. Phys.-Math. Soc. Japan **17** 48–57 [SPFNP 45–61]; 1939 On a Possible Interpretation of the Penetrating Component of the Cosmic Ray Proc. Phys.-Math. Soc. Japan **19** 712–713.
Yukawa H and Sakata S 1937 On the Interaction of Elementary Particles. II Proc. Phys.-Math. Soc. Japan **19** 1084–1093.
Yukawa H, Sakata S, and Taketani M 1938 On the Interaction of Elementary Particles. III Proc. Phys.-Math. Soc. Japan **20** 319–340.
Oppenheimer J R and Serber R 1937 Note on the Nature of the Cosmic-Ray Particles Phys. Rev. **51** 1113.

Surprise of European physicists
Kemmer N 1965 Problems on Fundamental Physics ed M Kobayasi (Kyoto: Kyoto) 602.

Lecture 11
Thomas precession
Møller C 1952 The Theory of Relativity (Oxford: Oxford University Press) § 22.
Thomas L H 1926 The Motion of the Spinning Electron Nature **117** 514; 1927 On the Kinematics of an Electron with an Axis Philosoph. Mag. **3** 1–22.

Relativistic wave equation for a particle with anomalous magnetic moment
Pauli W 1933 Die allgemeinen Prinzipien der Wellenmechanik Handbuch der Physik, vol 33 pt 1 ed H Geiger and K Scheel (Berlin: Springer) chap. 2, § 2, (esp. p 233)

[1980 trans Achuthan P and Venkatesan K *General Principles of Quantum Mechanics* (Berlin: Springer) chap. 9, § 18].

Derivation of Pauli's two-component equation from the Dirac equation
Pauli W Die allgemeinen Prinzipien der Wellenmechanik *Handbuch der Physik, vol 33 pt 1* ed H Geiger and K Scheel (Berlin: Springer) chap. 2, § 3, [esp. equation (89)] [1980 trans Achuthan P and K Venkatesan K *General Principles of Quantum Mechanics* (Berlin: Springer) chap. 9, § 19].

Doing the same for the anomalous magnetic moment and to compare it with the result of Thomas's theory
Jensen J H D 1972 verbally proposed to the author in Tokyo in September 1972.

Lecture 12
The content of lecture 12 is mostly from my memory. Therefore, I just mention the following, which I used to complement my memory.

Fierz M and Weisskopf V F ed 1960 *Theoretical Physics in the Twentieth Century: A Memorial Volume to Wolfgang Pauli* ed (New York: Interscience).

Goudsmit S. A. 1972 Guess Work: The Discovery of the Electron Spin *Delta* **15** 77–91.

Lectures by Heisenberg and Dirac 1932 "Various Problems in Quantum Theory," recording by Keimei Society **11** (translated by Yoshio Nishina). This last citation is a recording of the talks by Heisenberg and Dirac, who came to Japan in September 1929 and gave those lectures in Tokyo at the Institute of Physics and Chemistry. By the way, the Keimei Society is a foundation for the promotion of science established in 1918, its fund was one million yen, and the first director was Masatoshi Ohkohchi, who was the director of the Institute of Physics and Chemistry (RIKEN). This publication was sold widely for the price of 80 sen, but Mr. Ishikawa dug it up for the price of 700 yen [100 sen = 1 yen].

[Translator's Appended Bibliography
Since the original edition of this book was published in Japan in 1974, there have been several books published in English which cover related aspects of the history of physics of the period encompassed by this book. We found the following titles to be particularly enlightening and helpful while preparing this translation:

Brown L M, Pais A, and Pippard B ed 1995 *Twentieth Century Physics, Vol 1–3* (London: Institute of Physics and New York: American Institute of Physics)

Cahn R N and Goldhaber G 1989 *The Experimental Foundations of Particle Physics* (New York: Cambridge University Press)

Mehra J and Rechenberg H 1982–1987 *The Historical Development of Quantum Theory, Vol 1–5* (New York: Springer)

Pais A 1986 *Inward Bound: Of Matter and Forces in the Physical World* (New York: Oxford University Press)

Phillips M ed 1992 *The Life and Times of Modern Physics: History of Physics II: Readings from* Physics Today, *Number Five* (New York: American Institute of Physics)

Schweber S S 1994 *QED and the Men Who Made It: Dyson, Feynman, Schwinger, and Tomonaga* (Princeton: Princeton University Press)

Weart S R and Phillips M ed 1985 *History of Physics: Readings from* Physics Today, *Number 2* (New York: American Institute of Physics)

I also wish to bring to the reader's attention that the Japanese biography of Tomonaga is now available in English translation:

Matsui M and Ezawa H ed 1995; trans C Fujimoto and T Sano *Sin-itiro Tomonaga: Life of a Japanese Physicist* (Tokyo: MYU)]

INDEX